Easy Physics
STEP-BY-STEP

Easy Physics
STEP-BY-STEP

Jonathan S. Wolf

Mc
Graw
Hill
Education

New York Chicago San Francisco Athens London Madrid
Mexico City Milan New Delhi Singapore Sydney Toronto

1 2 3 4 5 6 7 8 9 10 QFR/QFR 1 0 9 8 7 6 5 4 3

ISBN   978-0-07-180591-9
MHID   0-07-180591-5

e-ISBN   978-0-07-180592-6
e-MHID   0-07-180592-3

Library of Congress Control Number 2012946831

Interior artwork by Progressive Information Technologies

Other titles in the series:
Easy Algebra Step-by-Step, Sandra Luna McCune and William D. Clark
Easy Biology Step-by-Step, Nichole Vivion
Easy Chemistry Step-by-Step, Marian DeWane
Easy Mathematics Step-by-Step, William D. Clark and Sandra Luna McCune
Easy Precalculus Step-by-Step, Carolyn C. Wheater

McGraw-Hill products are available at special quantity discounts to use as premiums and sales promotions or for use in corporate training programs. To contact a representative, please e-mail us at bulksales@mcgraw-hill.com.

This book is printed on acid-free paper.

Contents

Preface

With the wide variety of review books on the market designed to supplement coursework, you may be asking, "Why do we need another review book?" *Easy Physics Step-by-Step* is unique because it is *not* designed as a test preparation book. This book will guide you, step-by-step, through the material covered in a typical high school physics course. It will take a unique approach to problem solving without "teaching to the test." The emphasis is on presenting a review of fundamental concepts and encouraging you to think critically and creatively while solving problems.

In the spirit of this approach, Chapter 1 introduces a problem-solving ring as a guide to enhancing your skills as you progress through the material. You will notice that there are no multiple-choice questions. These types of assessments have their use, but the development of creative and critical thinking involves writing out the problems and their solutions in order to actively and constructively promote learning and understanding.

The material covered in this book is taught in most high schools. It reflects a beginning course in elementary physics and so tricky or difficult problems are kept to a minimum. Many sample problems are presented in a step-by-step manner to help ease you into solving more difficult problems. At the end of each chapter, there are several additional practice problems for you to solve on your own. An answer key is provided at the end of the book.

Topics and their level of difficulty have been selected based on 30 years' experience in teaching high school physics and represent material that will give the reader a good introduction and review of elementary physics. It is assumed that the reader has a working knowledge of basic algebra and trigonometry at the secondary school level.

I would like to thank Christopher Brown (from McGraw-Hill Education) and Grace Freedson (from the Grace Freedson Publishing Network) for facilitating this project. My colleagues Vanessa Blood, Robert Draper, Patricia Jablonowski, and Joseph Vaughan, from Scarsdale High School, have always been very supportive and helpful. Finally, I would like to thank my wife, Karen, my daughter Marissa and her fiancé Eli Lieberman, and my daughter Ilana for all of their love and support.

Easy Physics
STEP-BY-STEP

1

Introduction to Problem Solving in Physics

In this chapter you will learn about the methods of problem solving in physics. These are techniques that you will use throughout this book as you learn about the mechanical universe.

The Nature of Science

What is physics? This is a difficult question to answer. If you are reading this book, then you are either taking a class in physics or are interested in learning more about physics. Science presents us with a worldview that relies on our sense experiences in conjunction with the rules of logic. We observe the universe in a state of motion and change all around us. How do we make sense of it all? What framework and structure can we build to understand and make predictions about the universe and also understand the practical applications of this knowledge?

Science involves identifying problems; all problems have goals and givens. The task of a physicist is to identify the problems that can be solved using the so-called "scientific method." The method requires a scientist to identify the goals and givens in a particular problem.

As a physics student, you will be asked to solve problems that help you to understand some of the concepts covered in a typical high school physics class. To do so, you will use basic algebra and introductory trigonometry. To help you in this task, you will be introduced to step-by-step methods that will make the task of problem solving easier. These techniques will help you solve problems in other areas as well.

The Problem-Solving Ring and the Process Triangle

At the heart of any problem is the *goal*. Solving a problem involves first identifying the goal. The goal may be explicit: "Calculate the velocity of an object dropped from a height of 30 meters after it has been falling for 2 seconds." The goal may also be implicit, for example: "A person is driving a car at 25 m/s. When the car is 50 meters from an intersection, a traffic light turns from green to red. Does the car have enough time to stop?"

After identifying the goal of the problem, you need to evaluate all of the given information. What information is implied and what information is given explicitly? What general area of physics is the problem dealing with (motion, energy, electricity, optics, etc.)?

The next step is to decide on the best *solution path*. What concepts do you need to know? What equations will allow you to solve the problem efficiently? Once you have selected the path, you need to implement it. This usually involves substituting the given information into the equations and using the correct units (see Chapter 2). Before the chosen solution path can be correctly implemented, however, several steps may be required, as more

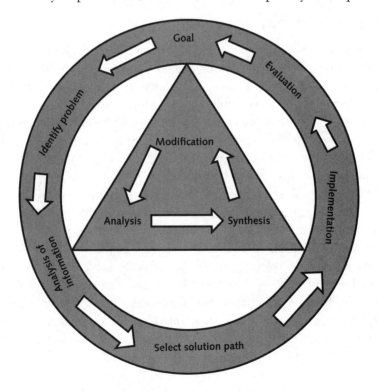

Figure 1.1 Problem-solving ring

information may need to be deduced. You may find your chosen solution path is either not correct or not the most efficient method (for example, given time constraints on an exam).

At the end of the mathematical process, you will obtain an answer. Now you must ask, "Does this answer make sense? Do the units correctly match the desired goal? Have I in fact reached the goal?" This is the evaluation stage. Once you evaluate your answer, you can decide if your goal or solution path needs to be modified. These steps are summarized in the *problem-solving ring* shown in Figure 1.1.

Embedded within this ring is a triangle representing three very important processes: analysis, synthesis, and modification. At all points along the problem-solving ring, you must continue to analyze (examine critically) your given information and problem-solving procedures. Synthesis involves linking together all of the different areas of physics and the problem-solving methods that help you to achieve the desired goal. Finally, you must be willing to modify what you are doing at every step. The triangle represents the idea that all three of these processes are linked and reinforce each other.

Whether you are solving a simple, one-step, "plug-and-chug" problem or a multistep problem, you can use this problem-solving cycle to help you. As you read through this book, you will see various aspects of this cycle employed. I have found it successful in the classes I teach, and I think you will find it successful as well.

2

Units and Measurements

In this chapter you will learn about how units and measurements are used in the study of physics. Then you will learn about how data is collected and analyzed using significant figures. This is a very important first step before we begin our exploration of the mechanical universe.

Fundamental (SI) Units

Physics is the study of the mechanical universe. Like all sciences, physics requires a set of rules and standards for making predictions about how the universe operates. You should be familiar with some of these rules, such as the scientific method, from your other science courses.

In physics, we use both observations and mathematical reasoning to establish relationships between physical quantities (like proportions). These proportions can also be visualized as patterns by using graphs (see Chapter 3).

We are also guided by the *rules of reasoning* outlined by Isaac Newton in his major work *Philosophiae Naturalis Principia Mathematica* (*The Mathematical Principles of Natural Philosophy*), published in 1687. They are summarized as follows.

1. Keep the number of causes to a minimum to explain an observed effect.

2. Assign the same cause to the same observed effect.

3. Certain universal qualities exist in all bodies (like length and mass).

4. Change our hypotheses about phenomena only when observed experimental evidence forces us to do so. If we do make those changes, then we must do more experiments to test the new hypotheses.

When making observations, we often use measurements as a way to establish a direct set of values for comparison. The idea of a *unit of measurement* allows us to make these comparisons fairly. The unit of measurement tells us not only what physical quantity we are measuring (length, area, volume, mass, energy, temperature, electric charge, frequency, etc.) but also what *scale* is being used for the measurement (we will discuss the concepts of *accuracy* and *precision* later in this chapter).

Problem   Which is larger, 500 millimeters (mm) or 5 meters (m)?

Solution

Step 1. The goal of this problem is to compare millimeters and meters. First, identify the type of measurements to be compared. In this case, we are comparing two lengths.

Step 2. For the solution path, identify the scale being used for comparison. In this case, the number 500 (called the magnitude) appears to be larger than the number 5. However, the scaling tells us that 1 meter = 1,000 millimeters, so 500 mm = 0.5 m, which is less than 5 m.

In physics, certain quantities are considered *fundamental* (recall Newton's fourth rule of reasoning) and, therefore, have *fundamental units*, also referred to as *Standard International* (or SI) units. They are summarized in Table 2.1.

The remaining units used in physics, called *derived units*, are all based on these fundamental units. For example, we will learn in Chapter 3 that average speed is a measure of distance divided by time. The derived units for this concept would be m/s. Area is measured in units of m^2, which is

Table 2.1   Fundamental SI Units

QUANTITY	NAME OF UNIT	ABBREVIATION
Length	Meter	m
Mass	Kilogram	kg
Time	Second	s
Electric current	Ampere	A
Temperature	Kelvin	K
Amount of substance	Mole	mol

derived from the units of length. As you learn a new concept, make sure you memorize the appropriate unit and use it in all calculations!

Conversion of Units

Sometimes quantities are not measured in fundamental units. In that case, we often have to convert them back into fundamental units. When recording measurements, certain prefixes are introduced to indicate the scale factor that is being used. Table 2.2 is a quick review of the powers of 10 and exponential notation. Table 2.3 shows some of the more common prefixes for units.

Table 2.2   Powers of Ten

$0.000000000000001 = 10^{-15}$	$1 = 10^{0}$
$0.000000000001 = 10^{-12}$	$10 = 10^{1}$
$0.000000001 = 10^{-9}$	$100 = 10^{2}$
$0.000001 = 10^{-6}$	$1000 = 10^{3}$
$0.001 = 10^{-3}$	$1000000 = 10^{6}$
$0.01 = 10^{-2}$	$1000000000 = 10^{9}$
$0.1 = 10^{-1}$	$1000000000000 = 10^{12}$
	$1000000000000000 = 10^{15}$

Table 2.3   Common Prefixes Used for Unit Conversions

PREFIX	ABBREVIATION	SCALE FACTOR	PREFIX	ABBREVIATION	SCALE FACTOR
Femto-	f	10^{-15}	Deka-	da	10^{1}
Pico-	p	10^{-12}	Hecto-	h	10^{2}
Nano-	n	10^{-9}	Kilo-	k	10^{3}
Micro-	μ	10^{-6}	Mega-	M	10^{6}
Milli-	m	10^{-3}	Giga-	G	10^{9}
Centi-	c	10^{-2}	Tera-	T	10^{12}
Deci-	d	10^{-1}	Peta-	P	10^{15}

Misconception

If you are using a calculator, you must not think that it is more precise to show a large number of digits after the decimal because the number of digits is an indication of the precision or accuracy of the measurements (refer to the section of "significant figures" later in this chapter). You must be very careful to make sure that the units you use are carefully matched and that the answer makes sense (recall the evaluation stage in the problem-solving ring). Knowing how to begin a problem is a big first step and it requires you to understand the units involved. Also, when proportions are used, you may sometimes see a problem stated in nonstandard units (like *atmospheres* for pressure instead of N/m^2). Given the context, this might be OK. In the end, standard units are always appropriate.

Problem Convert the following units (note that a *liter* [L] is a unit of *liquid volume*).

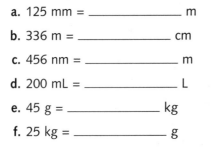

a. 125 mm = _____ m

b. 336 m = _____ cm

c. 456 nm = _____ m

d. 200 mL = _____ L

e. 45 g = _____ kg

f. 25 kg = _____ g

Solution

Step 1. The goal for these problems is to convert units using prefixes. For each conversion check the proportion multiplier.

a. 1 mm = 0.001 m

b. 1 m = 100 cm

c. 1 nm = 1×10^{-9} m

d. 1,000 mL = 1 L

e. 1 g = 0.001 kg

f. 1 kg = 1,000 g

Step 2. Perform the indicated conversion as needed.

a. 125 mm = 0.125 m

b. 336 m = 33,600 cm

c. 456 nm = 4.56×10^{-7} m

d. 200 mL = 0.200 L

e. 45 g = 0.045 kg

f. 25 kg = 25,000 g

Sometimes a problem will use common units for time that must be converted into seconds. Recall the following conversions.

1 microsecond (µs) = 10^{-6} seconds
1 millisecond (ms) = 0.001 seconds
60 seconds = 1 minute
60 minutes = 1 hour = 3,600 seconds
24 hours = 1 day (approximately) = 86,400 seconds
365.25 days = 1 year (approximately) = 31,557,600 seconds

Misconception

When you encounter a problem that involves time, it may be easier to work with units other than seconds. However, some of the formulas (and universal constants) may be valid only when the units of time are in seconds. You may see problems like this when studying orbits (where the orbital period of a planet might be expressed in earth-years) or other areas where something is presented in units of hours or minutes. The speedometer of a car may measure velocity (see Chapter 4) in km/hr instead of m/s. It is still appropriate to stay with standard units in all problems during calculations and then convert them as needed.

Problem Convert the following time measurements into seconds.

a. 300 ms = _____ s

b. 468 µs = _____ s

c. 18.5 minutes = _____ s

d. 2.5 years = _____ s

e. 5.77 days = _____ s

f. 36.5 hours = _____ s

Solution

Step 1. The goal for these problems is to convert all time units to seconds. For each conversion check the proportion multiplier.

a. 1 ms = 0.001 s

b. 1 µs = 10^{-6} s

  c. 1 minute = 60 s

  d. 1 year = 31,557,600 s

  e. 1 day = 86,400 s

  f. 1 hour = 3,600 s

Step 2. Perform the indicated conversion as needed.

  a. 300 ms = 0.300 s

  b. 468 μs = 4.68×10^{-4} s

  c. 15.5 minutes = 1,110 s

  d. 2.5 years = 78,894,000 s

  e. 5.77 days = 498,528 s

  f. 36.5 hours = 131,400 s

Accuracy and Precision

All measurements are subject to uncertainty. This does not mean that they are wrong! The choice of the correct instrument for measurement is related to the concepts of accuracy and precision. *Accuracy* refers to how close a measurement is to an accepted value for a physical quantity. *Precision* refers to the degree of agreement between successive measurements using a given instrument.

Let us suppose an experiment is performed to measure the acceleration due to gravity. As you will learn in Chapter 3, *acceleration* refers to a change in velocity (or speed) over time. It turns out that the accepted value for the acceleration due to gravity (near the surface of the earth) is the same for all dropped objects (objects that are falling from rest and with no air resistance acting on them to impede their motion). This is approximately equal to 9.81 m/s² (notice that the unit for acceleration is a *derived* unit; see Chapter 3 for more details). Suppose measurements were obtained by using a meterstick and a stopwatch to measure the time it takes objects to fall from various heights. The stopwatch can record hundredths of a second and the meterstick is marked with millimeters (0.001 m). Using formulas developed in Chapter 3, a student group obtains values of 8.36 m/s², 8.42 m/s², 8.38 m/s², and 8.39 m/s².

How would we analyze this data? In one sense, the students did not obtain the accepted value of 9.81 m/s². How consistent are their values? How reliable are their measurements? The values may appear to be consistent, but should the group have expected more accurate results given the measurement techniques? Should the students have obtained more data? If so, how much more?

These are difficult questions to answer. In reality, we must make a statement of what is an acceptable range of uncertainty for the purposes of accuracy. How much rounding did the students do? When they released the object, did the timer start immediately? Was it electronically controlled, or did someone say "go" and then someone else started the stopwatch by hand? What kind of data analysis should the students do? Should they use an average value? Should they make a graph of the data? How will they interpret their analysis? All of these issues should be discussed with the teacher before the experiment and in a laboratory report.

An average value (using a simple method of adding up the values and dividing by the number of values) would be equal to 8.3875 m/s² on a calculator. How is this number reported? Which numbers are significant? If the accepted value of the acceleration due to gravity is equal to 9.81 m/s², then maybe the students should report the average value as 8.39 m/s². A simple percent error, using the value of 8.39 m/s², could then be calculated from the formula

$$\% \text{ error} = \left| \frac{\text{experimental value} - \text{accepted value}}{\text{accepted value}} \right| \times 100$$

In this example, the percent error would be equal to approximately 14.5%. Is this an acceptable uncertainty? There are more sophisticated ways of analyzing data that we will not go into here. It is, however, important for you to understand that analysis of data is a crucial process in all areas of science.

 Misconception

Uncertainties are always present in science. This does not mean that your answer (or experimental procedure) is necessarily wrong. There may not be a "right" answer. In physics, there will be many times when approximations and limitations affect an answer. Conceptually, you may have to estimate using order of magnitude (powers of 10) approximations.

Significant Figures and Data Analysis

When a measurement is made, it is very important to inform the reader about the level of accuracy and precision used. Even if you are using a calculator to compute the solution to a problem, the number of digits written down is important. It is for this reason that we need to consider the issue of "significant figures."

The following sample problem shows you why the concept of significant figures is an important one.

Problem What is the length of the box if the ruler (not drawn to actual scale) records centimeters?

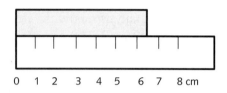

Solution

Step 1. The goal of the problem is to measure a length. Identify the scale divisions (smallest unit) used on the ruler. In this case, only centimeters are marked. It is not possible to precisely measure positions between the scaled intervals (for example, there are no marks to indicate tenths of a centimeter or smaller).

Step 2. Record the measurement of the length of the ruler to the lowest scaled marking. Here we can see that the length of the box is somewhere between 6 cm and 7 cm. The best we can say is that the box is at least 6 cm long. Anything else we report would be inaccurate and not precise. (Could we say with any certainty that the box measures 6.4 cm or 6.5 cm in length?) Our uncertainty in this case (worst estimate) is ±1 cm.

You should remember that all measurements are uncertain to some degree. It is important to report (since good experiments are reproducible) results as precisely and as accurately as possible. When a hypothesis is tested, it is subjected to scientific *peer review* and judged on the merits of the experiment's validity and reliability. This means that the experiment is judged on how careful the experimenter was when making and recording measurements. If the measurements are uncertain (or unreliable), then any mathematical analysis using those measurements will not only be uncertain, but sometimes magnified (for example, multiplying two measurements that are themselves uncertain by ±5% or more).

Notice that we could not say that the box in the problem is 6.0 cm long because we could not measure tenths of a centimeter. However, consider the next problem.

Problem Measure the length of the box shown using the ruler (not drawn to actual scale).

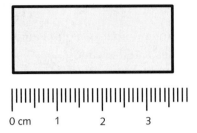

Solution

Step 1. The goal of the problem is to measure the length. Identify the scale divisions (smallest unit) used on the ruler. In this case, the ruler records measurements as small as 1 mm (one-tenth of a centimeter). We cannot precisely estimate (or measure) anything smaller than that.

Step 2. Record the measurement of the length of the box to the lowest scaled marking of the ruler. We can record the measurement as 3.6 cm (to the nearest mm). If the measurement were between 3.5 cm and 3.6 cm, we could not state the measurement precisely. In that case, our uncertainty would be about ±0.1 cm. If you wanted to convert cm to m, then the uncertainties would change as well (3.6 cm = .036 m and 0.1 cm = 0.001 m).

Using Significant Figures

The use of decimals and recording of uncertainties is part of reporting data. When recording (or reading) data, you need to be aware of the number of *significant figures*. The following general rules apply when using significant figures.

1. All numbers that are nonzero are significant.

2. Any zero that is between nonzero numbers is significant.

3. If there are any zeroes that are to the left of the first nonzero number, then those zeroes are not considered significant.

4. Zeroes that are to the right of a decimal point are significant.

5. If there are zeroes at the end of a number, but those zeroes are *not* to the right of a decimal point, then those zeroes may or may not be significant. This issue can be resolved using scientific notation (350 can be written as 3.5×10^2 or 3.50×10^2, as needed).

Important Tip

When rounding, be careful to follow guidelines established in your class for reporting significant figures.

Problem   State the number of significant figures in each of these numbers using the guidelines listed earlier.

   a. 3.684 kg

   b. 5.3 cm

   c. 209 L

   d. 5.02 m

   e. 0.045 m

   f. 0.0240 kg

   g. 1.20 × 10^5 m

Solution

Step 1. The goal of these problems is to identify the appropriate rule for significant figures.

Step 2. Determine the number of significant figures for each measurement.

   a. Four significant figures (all numbers are nonzero using rule 1).

   b. Two significant figures (all numbers are nonzero using rule 1).

   c. Three significant figures (the middle zero is significant by rule 2).

   d. Three significant figures (the middle zero is significant by rule 2).

   e. Two significant figures (the first zero to the left of the first nonzero number is not significant by rule 3).

   f. Three significant figures (the last zero, to the right of the decimal point is significant by rule 4, and the first zero to the left of the first nonzero number is not significant by rule 3).

   g. Three significant figures (without scientific notation, the number would have been 120,000 m. By writing the measurement as 1.20 × 10^5 m, there are 3 significant figures by rule 4).

Rule for Mathematical Operations

The accuracy of a mathematical operation is restricted to the least accurate measurement. For example:

a. 350.1 (4 significant figures) + 13.259 (5 significant figures) = 363.359, which must be reported as 363.3 (4 significant figures).

b. 1.2 (2 significant figures) × 5.65 (3 significant figures) = 6.78, which must be reported as 6.8 (2 significant figures).

Problem-Solving Strategies to Avoid Missteps

Refer to the problem-solving ring and process triangle as needed. Be very careful when rounding and make sure all units match the problem. In units of seconds, the number can get very large, so do not be confused if that occurs (one year is a long time when counted in seconds!). However, the prefixes (like kilo- or milli-) need to be converted back into standard units. Exercise 2.1 will allow you to test your understanding.

Exercise 2.1

1. Determine the number of significant figures in each of the following measurements.

   a. 9.354 kg

   b. 0.0670 m

   c. 22.4 L

   d. 25.0 m/s

For the remaining questions, perform the indicated operations and report your answers with the proper number of significant figures.

2. 3.4 + 13.2 + 123.25 =

3. 65.7 − 9.43 =

4. 3.4 × 6.45 =

5. 12.2/2 =

6. $(4.5)^2$ =

7. Find the average value of 3.5 m, 3.6 m, 3.4 m, 3.5 m, 3.7 m, and 3.8 m.

3

Graphical Analysis of Data

In this chapter you will learn to make simple graphs, using proper graphing techniques that will represent different physical relationships. These proportional relationships will then provide the starting points for the different equations that will be developed as we progress in our exploration of the mechanical universe.

How to Make a Simple Linear Graph

When making any graph, it is important to remember these steps.

1. Every graph has a horizontal (x) axis and a vertical (y) axis.

2. The independent quantity (the quantity that is varied in an experiment) is usually plotted on the horizontal axis, and the dependent quantity (the quantity that is affected by the independent quantity) is usually plotted on the vertical axis. Sometimes, the quantity of time is plotted on the horizontal axis to measure a *rate of change*.

3. Both axes should be labeled with the names of the quantities and their associated units. The graph should also have a title.

4. When graphing by hand, an appropriate scale should be selected. The scaled axes should begin at the origin (0,0), but that point may or may not be plotted (depending on the needs of the individual experiment). The scales should be uniform and may or may not reflect actual data values.

5. Points are plotted as precisely as possible (recognizing the inherent uncertainty in measured values). If a plotted point needs to be estimated (or *interpolated*), then care must be taken to use grid lines that allow for a minimum of uncertainty.

6. For a pattern that appears to be linear, a *best-fit line* is drawn that represents the *trend line* (this is sometimes called a *regression line*). The regression line may or may not go through all of your data points. For other patterns, best-fitting smooth curves should be drawn. All subsequent analyses will be based on these best-fit curves.

Problem   A student observes a toy car moving along the floor. When the car passes a given point (designated as the origin), the student begins to use a stopwatch (calibrated to read hundredths of a second) to record the time it takes the car to pass distances previously marked using a meterstick. The data is recorded as follows:

TIME (S)	DISTANCE (M)
0	0
0.48	1.0
0.97	2.0
1.51	3.0
1.98	4.0
2.52	5.0
3.02	6.0

Plot the graph of distance versus time for this data on the grid.

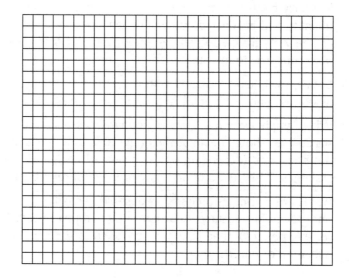

Solution

Step 1. Label the axes (notice that time is plotted on the *x* axis because we will be comparing the rate of change of distance with time). Make sure you include the units.

Step 2. Pick an appropriate scale for each axis and mark these points on the graph. The point (0,0) is included in this graph because the observations began at the origin (arbitrarily chosen).

Step 3. Plot the points and interpolate as needed.

Step 4. Draw a regression (best-fit) line through the points. Give a title to the graph.

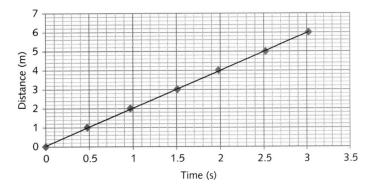

Misconception

When making graphs by hand, be sure to pick an appropriate scale and size so that the graph can be easily analyzed. If an analysis requires you to identify points other than those plotted, and if they are between the given data points, that procedure is called *interpolation*. If the points to be analyzed lie outside the graph, but along the trend line, then the process is called *extrapolation*. It is important not to use data points as a scale unless they are equally spaced.

This graph of distance versus time is an example of a *linear direct relationship*. We would say that distance varies directly with time (the *linearity* would be implied, but not always stated). The graph represents the equation of a line that begins at the origin and is sometimes written (in physics) as

$$y = kx$$

where *k* is a constant (in this case, the slope of the line).

The slope of a line is given by

$$k = \frac{\Delta y}{\Delta x} = \frac{y_2 - y_1}{x_2 - x_1}$$

where the symbol Δ means "change" or "difference."

In mathematics, the equation of a line is sometimes written as $y = mx$. However, in physics, the letter "m" can be confused with the unit "meter" or represent the concept "mass." To avoid this, physicists will use the letter k to represent a generic constant. In the case of a line, the constant k corresponds to the slope of the line.

If there is a y-intercept, then the equation would be written as $y = kx + b$. In physics, all of these symbols have meanings and some of them (not necessarily all) usually have units associated with them.

In our example, the slope of the line corresponds to the rate of change of distance with time (which will be defined as *speed* in Chapter 4). To evaluate the slope, simply work from the trend line (not the individual data points, because they may not lie on the best-fit line due to experimental uncertainties). Using the best-fit line, the value of the slope is $k = 2.0$ m/s. In Chapter 4, you will learn that this slope corresponds to the *average speed* of the motion.

Proportional Relationships

There are several other important proportional relationships that commonly appear in most physics problems.

Problem A student measures the distance traveled by a moving object. The data collected required the student to observe the distance of the object at regular intervals of time. Using proper graphing techniques, make a graph of the data as shown in the table.

TIME (S)	DISTANCE (M)
0	0
1.0	3.0
2.0	12.0
3.0	27.0
4.0	48.0
5.0	75.0
6.0	108.0

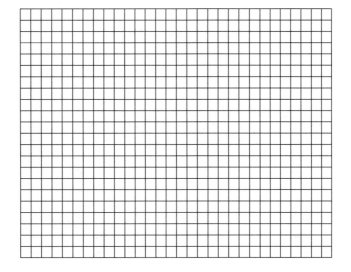

Solution

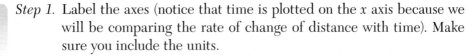

Step 1. Label the axes (notice that time is plotted on the *x* axis because we will be comparing the rate of change of distance with time). Make sure you include the units.

Step 2. Pick an appropriate scale for each axis and mark these points on the graph. The point (0,0) is included in this graph because the observations began at the origin (arbitrarily chosen).

Step 3. Plot the points and interpolate as needed.

Step 4. Draw a best-fit curve if the pattern does not appear to be linear.

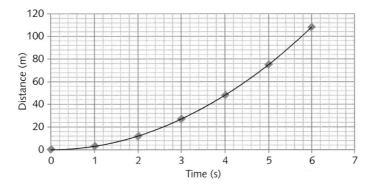

This graph of distance versus time is an example of a simple *quadratic (squared) relationship* (starting from the origin). It is not linear, so we would say that distance varies directly with the square of time. The general equation for this relationship can be written as

$$y = kx^2$$

where k is a proportionality constant.

This is the equation for a *parabola* and the value (or *magnitude*) of the constant can be determined by noticing that for this relationship

$$k = y/x^2 = 3 \text{ m/s}^2$$

For reasons that you will learn in Chapter 4, this constant is related to the *acceleration* of the object (but it is not equal to the actual acceleration!).

Problem From introductory chemistry, it is known that a relationship exists between the pressure of a confined gas and its volume (if the temperature remains constant). This is known as *Boyle's Law*.
A student obtains the following data in an experiment. Note that the gas pressure is measured in units of atmospheres (atm) and the volume is measured in units of liters. Both of these units are derived units (recall Chapter 2). Make a graph of the data in the table using proper graphing techniques.

PRESSURE (ATM)	VOLUME (LITERS)
3.0	15.0
2.5	18.0
2.0	22.5
1.5	30.0
0.5	45.0
0.3	90.0
6.0	150.0

Solution

Step 1. Label the axes. Make sure you include the units.

Step 2. Pick an appropriate scale for each axis and mark these points on the graph.

Step 3. Plot the points and interpolate as needed.

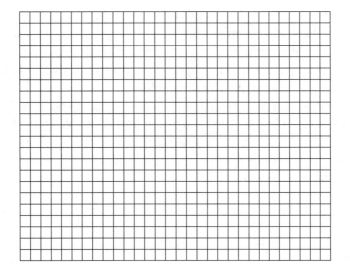

Step 4. Draw a best-fit curve if the pattern does not appear to be linear.

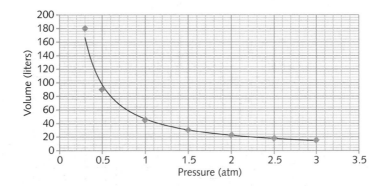

This graph of pressure versus volume is an example of an *inverse relationship*. We would say in an inverse relationship that volume varies inversely with pressure. The general equation for this relationship is expressed as

$$xy = k \quad \text{or} \quad y = k/x$$

where k is equal to a constant.

This is the equation of a *hyperbola*. The magnitude of the value of k can be found directly from the relationship $k = xy$. In this case, the numerical value is $k = 3$. (We will not discuss the meaning of the units for k here; you can refer to any chemistry book for more details).

Problem   A simple pendulum consists of a weight tied to a string that is then suspended from a support. As the pendulum swings, the time to make one complete cycle (back and forth) is called the *period* (you will study this concept further in Chapter 7). A student performed an experiment in which the period of a simple pendulum was observed to vary as the length of the pendulum varied. The following data was collected. Using proper graphing techniques, make a graph of the data given in the table.

LENGTH (M)	PERIOD (S)
0	0.0
0.20	0.89
0.30	1.10
0.40	1.26
0.50	1.41
0.60	1.55
0.70	1.67

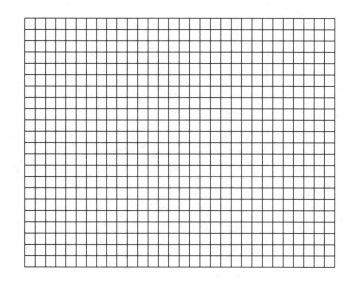

Solution

Step 1. Label the axes. Make sure you include the units.

Step 2. Pick an appropriate scale for each axis and mark these points on the graph.

Step 3. Plot the points and interpolate as needed. In this case, the point (0,0) is included.

Step 4. Draw a best-fit curve if the pattern does not appear to be linear.

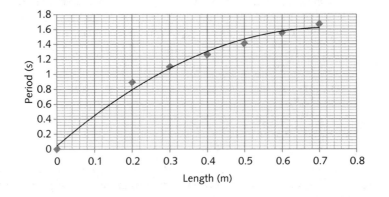

This is an example of what is called a *direct square root relationship*. The graph is in the shape of a sideways parabola. In this case, we would say that the period of a simple pendulum varies directly with the square root of its length. The general equation for this relationship is given by

$$y = k\sqrt{x}$$

where k = constant. The value (neglecting the units for now) of k can be found from the relationship

$$k = \frac{y}{\sqrt{x}} = 2.00 \ (\text{approximately})$$

The main relationships we have discussed are summarized in the following list. It will be important for you to remember them (and their graphs, as presented earlier) as we begin our journey of exploration of the mechanical universe.

Summary of Common Relationships in Physics

Linear direct relationship	$y = kx$
Direct squared (quadratic) relationship	$y = kx^2$
Inverse relationship	$y = k/x$
Inverse square relationship	$y = k/x^2$
Direct square root relationship	$y = k\sqrt{x}$

Problem-Solving Strategies to Avoid Missteps

Refer to the problem-solving ring as needed. Make sure that you test each relationship rather than making an assumption beforehand. Even if you will be using a computer or a graphing calculator, make sure you understand the parameters that are being used. Trend lines (or best-fit curves) can lead to errors, but they should be used for analysis (see Chapter 4 regarding slopes, tangents, and areas under graphs). Exercise 3.1 will give you a chance to test your understanding.

Exercise 3.1

For each set of data, make a proper graph, state the relationship in words, find the value of the constant k, and write the final equation. There are no units for the simulated data presented.

1.

x	y
0	5
0.5	7
1.0	9
1.5	11
2.0	13
2.5	15

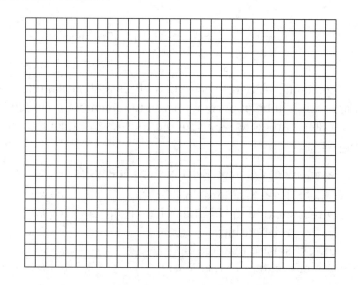

2.

x	Y
10	2.00
20	1.00
30	0.67
40	0.50
50	0.40
60	0.33

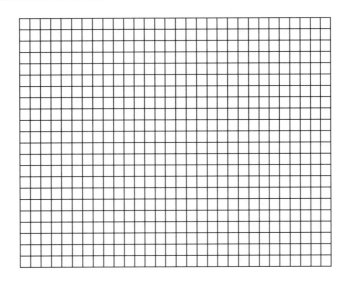

3.

x	Y
0	0
2.0	12
4.0	48
6.0	72
8.0	192
10.0	300

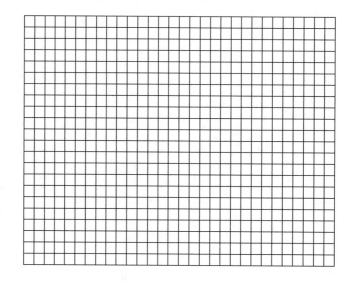

4.

x	Y
0	0
3	8.66
6	12.25
9	15.00
12	17.32
15	19.36

4

Linear Motion

In this chapter you will learn by using graphical analysis and algebraic equations how to describe the motion of an object moving in one dimension, a concept called *kinematics*. Because everything in the universe is in motion, an understanding of kinematics is an essential step on our journey of exploration in physics.

Motion and Frame of Reference

All observations are based on the point of origin (viewpoint) of the observer. This is called a *frame of reference*. The frame of reference identifies the location (placement) of the object. *Motion* is defined to be the observed change in position of an object relative to (this means "based upon") a given frame of reference. This change in position is called *displacement*. For example, if you hold a book in your hand while you are walking, the book is at rest relative to you but is observed to be in motion relative to someone sitting on a chair watching you.

Scalar

Any quantity that is described only by its *magnitude* (or size) is called a *scalar*. Mass, length, area, volume, distance, and speed are all examples of scalars.

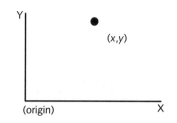

Figure 4.1 A frame of reference

Vector

Any quantity that is described by its magnitude *and* direction (relative to a frame of reference) is called a *vector*. Displacement, velocity, acceleration, force, and momentum are all examples of vectors.

Misconception

The concept of a frame of reference can be confusing. The selection of a frame of reference is arbitrary, but we usually think of the observer as being the origin of a coordinate system. We identify a position, relative to an origin, by setting up a set of two or three coordinates (x,y) or (x,y,z), depending on how many dimensions we are considering. This is illustrated in Figure 4.1.

It is also very important to remember the differences between vector quantities and scalar quantities (more details will follow in Chapter 5). The differences between distance and displacement are just a first example.

Kinematic Quantities

Distance is a scalar quantity measuring the length between two positions in units of meters. Distance is usually designated by a lowercase (light typeface) *d* (to indicate that it is a scalar).

Displacement is a vector quantity (also measured in units of meters) that represents the change in position of an object (relative to a given frame of reference). We will designate displacement by a bold typeface letter **d** (in most cases vector quantities will be displayed in a bold typeface). In other words, the magnitude of the displacement is the distance!

Displacement is measured from the starting point of the motion directly to the ending point of the motion. For example, if I want to go from point A to point B, along a curved path, the final displacement is just the direct arrow drawn from point A to point B (see Figure 4.2).

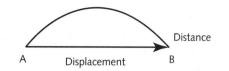

Figure 4.2 Displacement

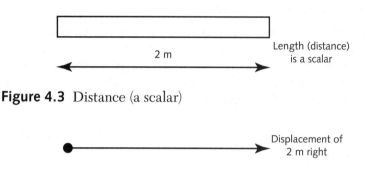

Figure 4.3 Distance (a scalar)

Figure 4.4 Displacement vector

As another example, let us imagine that the length of a box is 2 meters (Figure 4.3). This distance is the same whether I measure it from left to right or from right to left. If I am told an object has been moved a distance of 2 meters, I do not know where it is now, since it could have been moved 2 meters to the right or 2 meters to the left (or in any other direction relative to myself or to another person). However, if I am told that an object has been moved 2 meters to my right (Figure 4.4), then I know exactly where it is now located. (Please note that none of these diagrams is drawn to scale.)

Problem   A girl walks 8 m north and then 3 m south. What distance did she travel? What is her final displacement?

Solution

Step 1. Find the distance by recognizing that distance is a scalar and the total accumulated length, in this case, is $d = 11$ m.

Step 2. Find the final displacement by recognizing that the directions north and south are opposite to each other (think of north as + and south as –). Relative to the starting point, she first started from her origin, walked 8 meters north, and then walked 3 meters in the opposite direction. She has ended her motion 5 m north of her origin.

Important Tip

When combining displacements that are in the *same* direction, you would *add* their magnitudes. When combining displacements that are in the *opposite* direction, you would *subtract* their magnitudes. This will be true for *all* vectors. The combination of two or more vectors is called the *resultant*.

Misconception

The addition or combination of vectors is not the same as simply adding or subtracting numbers. Vectors (as we shall see in Chapter 5), can be combined in two or three dimensions, which will require the use of trigonometry. In these simple examples, the vectors are one-dimensional and directional changes are simply designated as positive (+) or negative (−).

The average *speed* (scalar) of an object is equal to the total distance divided by the total time.

$$v_{avg} = \frac{\Delta d}{\Delta t} \text{(units are m/s)}$$

where Δ means *change*.

The *average velocity* (vector) of an object is equal to the rate of change of its *displacement*:

$$\mathbf{v}_{avg} = \frac{\Delta \mathbf{d}}{\Delta t} \left(\text{units are also m/s, but with a direction}\right)$$

The *average acceleration* (vector) of an object is equal to the rate of change of its *velocity*:

$$\mathbf{a} = \frac{\Delta \mathbf{v}}{\Delta t} \left(\text{units are m/s}^2\right) \text{ and } \Delta \mathbf{v} = \mathbf{v}_f - \mathbf{v}_i = \text{final velocity} - \text{initial velocity}$$

We can also write that $\Delta \mathbf{d} = \mathbf{v}_{avg} \Delta t$ or more simply $\mathbf{d} = \mathbf{v}_{avg} t$ for constant velocity.

Important Tip

You must remember that velocity and speed are *not* the same concept! The *acceleration* is a vector quantity equal to the *change* in velocity. This means that the object may be changing its speed and/or changing its direction. The units of acceleration are m/s² and are read as "meters per second squared" or "meters per second per second."

The equation for acceleration can also be written in the following form:

$$a = \frac{\mathbf{v}_f - \mathbf{v}_i}{t}$$

and

$$\mathbf{v}_f = \mathbf{v}_i + \mathbf{a}t$$

This final velocity is also known as the *instantaneous* velocity.

Important Tip

You must remember that in linear motion, the velocity and acceleration can have both + or − values. You should also remember to include units in all substitutions and final answers. You must always use fundamental units (*displacement* must be in meters; *time* must in seconds)!

Problem   An object is traveling with a constant velocity of 10 m/s east. How far will it have traveled after 3 minutes?

Solution

Step 1. Since the velocity is constant, understand that you can use the formula $\mathbf{d} = \mathbf{v}t$.

Step 2. The time needs to be converted from minutes to seconds (required): 3 minutes = 180 s.

Step 3. Substitute your values into the formula with units (the displacement is our unknown) and solve: $\mathbf{d}$ = (10 m/s)(180 s) = **1800 m east**.

Misconception

It is *not* correct to assume that an object with zero acceleration is not moving in a given frame of reference. Zero acceleration simply means that the velocity is not changing! Constant velocity could mean that the object is either not moving or is at rest. An object can also have a zero average *velocity* while having a nonzero average *speed* (if it is going back and forth). A change in velocity could mean a change in speed and/or a change in direction. Also, since it is a vector, acceleration can be designated as positive (+) or negative (−). However, if we define positive velocity as forward motion (and thus negative velocity implies backward motion), the sign of the acceleration tells us the direction of the *change* in the velocity. This means that a *positive* acceleration could mean one of two things.

1. An object is *speeding up* in the *forward* direction.
2. An object is *slowing down* in the *backward* direction.

A similar set of ideas would apply to the meaning of negative acceleration (which is *not* always used to mean the more commonly known concept of slowing down, or *deceleration*).

Important Tip

For most physics problems, it is important to identify and write down all of the given information. You can either make a chart or just list the information neatly (include units!). Refer back to the problem-solving ring in Chapter 1! Next, identify the appropriate formula to use based on the given information. Write the formula and substitute your given values (with units). Use algebraic techniques to solve for the missing quantity. In the earlier problem, v = constant = 10 m/s east and t = 3 minutes = 180 s.

Problem   An object has an initial velocity of 5 m/s east and then speeds up to 25 m/s east in 4 s. What is the acceleration of the mass?

Solution

Step 1. The goal of this problem is to find the value of the acceleration. First, identify the given information.

v_i = 5 m/s , v_f = 25 m/s, t = 4 s (we will select the direction *east* as +)

Step 2. Identify the appropriate equation and substitute given values (with units):

$$\mathbf{a} = \frac{\mathbf{v}_f - \mathbf{v}_i}{t} = \frac{25 \text{ m/s} - 5 \text{ m/s}}{4 \text{ s}}$$

$$= +5 \text{ m/s}^2 \text{ east (same direction as both velocities)}$$

Problem   An object is moving with a velocity of 50 m/s and slows to 10 m/s in 10 s. What is the acceleration of the object?

Solution

Step 1. The goal of this problem is to find the acceleration. First, identify the given information:

v_i = 50 m/s, v_f = 10 m/s, t = 10 s

Step 2. Identify the appropriate equation and substitute the given values
(with units) into it.

$$\mathbf{a} = \frac{\mathbf{v}_f - \mathbf{v}_i}{t} = \frac{10 \text{ m/s} - 50 \text{ m/s}}{10 \text{ s}} = -4 \text{ m/s}^2 \text{ (west)}$$

Important Tip

Notice that even though the object is still moving *forward* (east, or +), the
direction of the acceleration is *west* (–) since the object is slowing down!

Problem An object is moving north with a velocity of 12 m/s. It accelerates at a rate of 3 m/s² for 10 s. What is the velocity of the object after that
time?

Solution

Step 1. The goal of this problem is to find the instantaneous final velocity.
First, identify the given information:

$\mathbf{v}_i$ = 12 m/s, $\mathbf{a}$ = 3 m/s², t = 10 s

Step 2. Identify the appropriate equation and substitute in the given values
(with units).

$\mathbf{v}_f = \mathbf{v}_i + \mathbf{a}t = (12 \text{ m/s}) + (3 \text{ m/s}^2)(10 \text{ s}) = 42 \text{ m/s north}$

Finding Displacement with Uniformly Accelerated Motion

If the velocity is not constant, then the object is accelerating by either speeding up, slowing down, and/or changing direction. If the acceleration is constant, then the average velocity is also given by the equation:

$$\mathbf{v}_{avg} = \frac{\mathbf{v}_i + \mathbf{v}_f}{2}$$

This means that our original equation for displacement can now be written as:

$$\mathbf{d} = \left(\frac{\mathbf{v}_i + \mathbf{v}_f}{2} \right) t$$

If we substitute the formula for v_f into the earlier equation for displacement, we obtain:

$$d = v_i t + \frac{1}{2} at^2$$

Finally, if we use the equation for the definition of acceleration and solve for the time,

$$t = \frac{v_f - v_i}{a}$$

and then substitute it for t in the equation for displacement and average velocity, we get:

$$d = \left(\frac{v_f + v_i}{2} \right) \left(\frac{v_f - v_i}{a} \right) = \frac{v_f^2 - v_i^2}{2a}$$

or

$$v_f^2 - v_i^2 = 2ad$$

Important Tip

These kinematics equations are only valid when the acceleration is *uniform*, or constant.

Problem An airplane accelerates from rest to a speed of 125 m/s while traveling a distance of 500 m. What is the acceleration of the plane? How long did it take for the plane to reach that velocity?

Solution

Step 1. The goal of this problem is to find the time of travel. First, identify the given information:

$v_i = 0$, $v_f = 125$ m/s, $d = 500$ m (note that the time is not given!)

Step 2. Identify the appropriate equation and substitute given values (with units). Since the time is not given, then it is appropriate to use the equation

$v_f^2 - v_i^2 = 2ad$

$(125 \text{ m/s})^2 - 0 = 2a(500 \text{ m})$

which means

a = 15.63 m/s²

Step 3. Now use the value of the acceleration and the velocities to calculate the time:

$$\mathbf{a} = \frac{\mathbf{v}_f - \mathbf{v}_i}{t} \text{ or } t = \frac{\mathbf{v}_f - \mathbf{v}_i}{\mathbf{a}} = \frac{125 \text{ m/s} - 0}{15.63 \text{ m/s}^2} = 8 \text{ s}$$

Freely Falling Objects

Galileo Galilei (1564–1642) was the first scientist to determine the acceleration of a falling object. He demonstrated that all objects fall with the same acceleration in the absence of any external forces (such as air resistance). In modern units, this acceleration is given by:

$$\mathbf{g} = -9.8 \text{ m/s}^2 \text{ (in the downward direction)}$$

Important Tip

For all vertically moving objects that are under the influence of gravity only, the value of the acceleration is already known to be equal to **g**. This is true whether the object is thrown upward, thrown downward, or simply dropped from rest. With no air resistance, two objects that are dropped from the same height simultaneously will hit the ground at the same time!

Problem A ball is dropped from an unknown height. It is observed to hit the ground in 3.5 s. From what height was it dropped (ignore air resistance)? How fast will it be going as it hits the ground?

Solution

Step 1. There are two goals in this problem. First, we need to find the displacement of the object. Second, we need to find the velocity as it hits the ground. However, the phrase "how fast" might mean that all we need to do is find the speed (magnitude of the velocity). First, identify the given information.

In this problem, "dropped" is a key code word. It implies that the initial velocity of the ball was zero and that the acceleration will be equal to **g**.

$$\mathbf{a} = \mathbf{g} = -9.8 \text{ m/s}^2, t = 3.5 \text{ s}$$

Step 2. Identify the appropriate equation and substitute in the given values (with units). Since $\mathbf{v}_i = 0$ and we do not know the final velocity as it hits, we can use the equation

$$\mathbf{d} = \frac{1}{2}\mathbf{a}t^2 = (0.5)\left(-9.8 \text{ m/s}^2\right)(3.5 \text{ s})^2 = -60.0 \text{ m}$$

Step 3. We need to report the answer correctly for the first part. Notice that our answer is *negative*. This is because the equation solves for the *displacement*, which is a *vector* quantity. If the question asks for the displacement, then the ball has fallen approximately 60 m in the downward direction (relative to the starting point). If the question simply asks for the height, then this is a *scalar* quantity (the magnitude of the displacement), and we can report the height as equal to approximately 60 m.

Step 4. Identify the appropriate equation for the second part of the question and substitute in the correct values (with units):

$\mathbf{a} = \mathbf{g} = -9.8 \text{ m/s}^2, \mathbf{v}_i = 0, t = 3.5 \text{ s}$

Use

$$\mathbf{v}_f = \mathbf{v}_i + \mathbf{a}t = 0 + (-9.8 \text{ m/s}^2)(3.5 \text{ s}) = -34.3 \text{ m/s}$$

Again, notice that the final velocity is negative since the object is moving downward and velocity is a *vector* quantity. The speed (which is a *scalar* quantity) would be stated as 34.3 m/s.

Misconception

The acceleration due to gravity is always equal to -9.8 m/s^2 whether an object is being thrown upward or downward (as long as no other forces are acting upon it; see Chapter 6). When an object thrown upward reaches its maximum height and momentarily *stops* (to change direction), the acceleration of the object is *not* equal to zero at that point!

Problem An object thrown upward rises to a maximum height, stops, turns around, and then falls back again. It takes the same amount of time to rise as it takes to fall (through the same distances). Suppose a ball is thrown upward with an initial velocity of +25 m/s. How long will it take until it reaches its maximum height? What is the magnitude of the maximum height? What will be the velocity when it lands back into the hands of the thrower (assuming it fell back down through the same height)?

Solution

Step 1. There are three goals in this problem. We are asked to find the time it takes the object to rise to its maximum height, the magnitude of the maximum height (or its displacement relative to the ground), and finally, what the velocity will be as it returns to the original position. First, we need to identify the given information.

$$\mathbf{v}_i = 25 \text{ m/s}, \mathbf{v}_f = 0, \mathbf{a} = \mathbf{g} = -9.8 \text{ m/s}^2$$

Notice that the acceleration is $\mathbf{g}$ (which is implied) and is still *negative* even though the object is moving upward initially. This is because gravity is a *force* (a push or a pull; see Chapter 6) that is slowing down the ball as it rises. When the ball is falling, gravity is accelerating it downward, making it go faster in the downward direction (so velocities would be negative in that case!). Also notice that the *instantaneous velocity* is equal to zero at the maximum height!

Step 2. Identify the appropriate equation and substitute in the given values (with units). To find the time to reach the maximum height, we can use the formula for acceleration:

$$\mathbf{a} = \frac{\mathbf{v}_f - \mathbf{v}_i}{t}$$

and solve for time.

$$t = \frac{\mathbf{v}_f - \mathbf{v}_i}{\mathbf{a}} = \frac{0 - 25.0 \text{ m/s}}{-9.80 \text{ m/s}^2} = 2.55 \text{ s}$$

This is the time it takes gravity to *decelerate* the object to a velocity of zero at the maximum height. Be careful to make sure that you do not end up with a negative time (which is physically meaningless!).

Step 3. Identify the appropriate equation to find the maximum height and substitute in the appropriate values (with units). We have several choices for the equations. Since $\mathbf{v}_i = +25$ m/s, $\mathbf{v}_f = 0$, acceleration is constant ($\mathbf{g}$), and $t_{up} = 2.55$ s, we can simply use:

$$\mathbf{d} = \left(\frac{\mathbf{v}_i + \mathbf{v}_f}{2}\right)t = \left(\frac{25 \text{ m/s} + 0}{2}\right)(2.55 \text{ s}) = 31.88 \text{ m} = 32 \text{ m}$$

Graphical Analysis of Linear Motion

The motion of objects moving in one dimension can be analyzed using graphs. For example, if an object starts from the origin and moves forward with constant velocity, then there exists a direct relationship between displacement and time (Figure 4.5).

The slope of the graph of displacement versus time is equal to the average velocity of the motion. The sign of the slope is important, because it corresponds to the direction of motion.

If an object is at rest, then there is no motion (in a given frame of reference) and the displacement remains constant over time (Figure 4.6). The slope (*average velocity*) is equal to *zero* in this case.

The *magnitude* of the slope of the graph would be equal to the *average speed*. Recall that *average speed* is just the ratio of the total distance divided by the total time. It is also possible to have a situation where the *average velocity* is equal to zero but the *average speed* is not! If an object goes away from the origin, turns around, and comes back, its final displacement is zero even though it traveled a certain total distance. The *average velocity* in this case is equal to zero.

If an object were to go backward from a starting point past the origin at a constant velocity, then the relationship between displacement and time would still be direct and linear but with a negative slope (Figure 4.7).

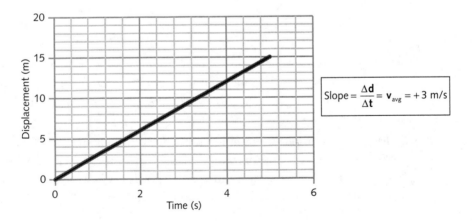

$$\text{Slope} = \frac{\Delta d}{\Delta t} = v_{avg} = +3 \text{ m/s}$$

Figure 4.5 Relationship between displacement and time: constant velocity

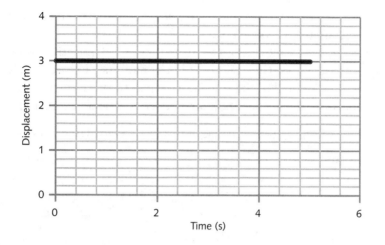

Figure 4.6 Relationship between displacement and time: object at rest

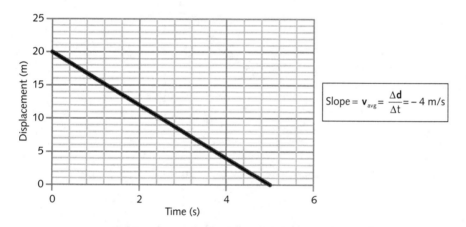

Figure 4.7 Relationship between displacement and time: negative velocity

If an object undergoes positive uniform acceleration from rest (starting from the origin), then the relationship between displacement and time is quadratic (since $\mathbf{d} = \frac{1}{2}\mathbf{a}t^2$). The slope of a tangent line to the graph would represent the *instantaneous* velocity at that time (Figure 4.8). In the figure, a tangent line is drawn at 2.4 s. The slope of the tangent line is approximately equal to +10 m/s (corresponding to the instantaneous velocity at

2.4 s). Notice that tangent lines are *not* exact and are only an approximation of the instantaneous velocity.

If an object is moving forward with constant (positive) velocity, then a graph of velocity vs. time will be a horizontal line (Figure 4.9). The average acceleration is equal to the slope of the line. In this case, because the velocity is constant, the acceleration (and hence the slope) is equal to zero.

Area of rectangle = change in displacement = (10 m/s)(4 s) = 40 m

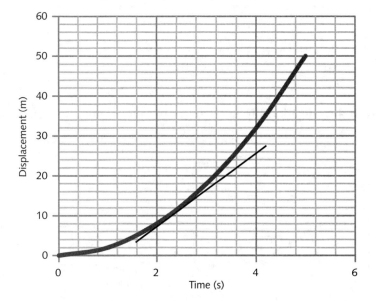

Figure 4.8 Instantaneous velocity measured by tangent line

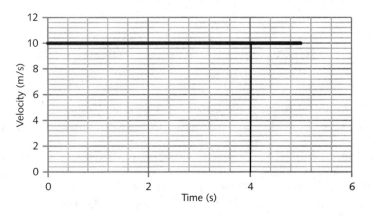

Figure 4.9 Constant velocity: Finding displacement as the area under a velocity versus time graph

Important Tip

The area underneath a velocity vs. time graph represents the change in displacement of the object during a specified interval of time. In the earlier example

If an object has uniform positive acceleration from rest, then a graph of velocity vs. time will be a direct linear relationship (Figure 4.10). The slope of the line in Figure 4.10 is equal to 2 m/s² and represents the *average acceleration* for this motion (which started from *rest*).

The change in displacement is given by:

$$\text{Area of triangle} = \text{change in displacement} = \frac{1}{2}\left(\text{base} \times \text{height}\right)$$

$$= \frac{1}{2}(8 \text{ m/s})(4 \text{ s}) = 16 \text{ m}$$

Important Tip

Notice that in Figure 4.10 the *average* velocity is equal to 4 m/s. We could then calculate the change in displacement by using:

d = v*t* = (4 m/s)(4 s) = 16 m

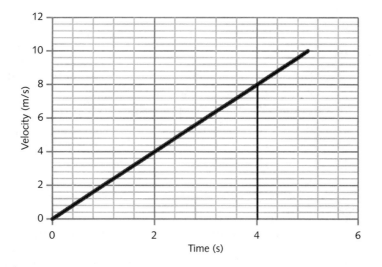

Figure 4.10 Constant acceleration: Finding displacement as the area under a velocity versus time graph

Misconception

The area under a velocity vs. time graph is *not* equal to the displacement, but equal to the *change* in the displacement. If the object starts from the origin, then the displacement is compared to a zero starting point.

Problem   A graph of velocity vs. time for a car is shown. Find:

a. The acceleration of the object from $t = 0$ s to $t = 20$ s

b. The acceleration of the object from $t = 20$ s to $t = 60$ s

c. The acceleration of the object from $t = 60$ s to $t = 70$ s

d. The total displacement traveled by the object during the entire 70 s

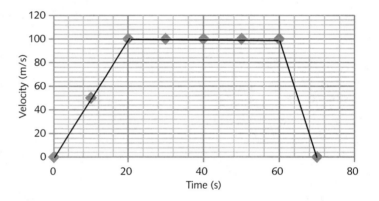

Solution

Step 1. The goals of this problem are to find the value of the acceleration of the object at various time intervals and to find the final displacement of the object starting from rest. First, recall that the slope of the line, in a velocity vs. time graph, represents the *average* acceleration. Now, find the slope of the line in each time interval stated:

a. Slope $= \mathbf{a} = \dfrac{100 \text{ m/s} - 0}{20 \text{ s}} = +5 \text{ m/s}^2$

b. Slope $= \mathbf{a} = 0$ m/s² (since the line is horizontal from $t = 20$ s to $t = 60$ s)

c. Slope $= \mathbf{a} = \dfrac{0 - 100 \text{ m/s}}{10 \text{ s}} = -10$ m/s² (object is slowing down, while going forward)

Step 2. To find the total displacement from $t = 0$ s to $t = 0$ s, we will divide the area into three segments of rectangles and triangles.

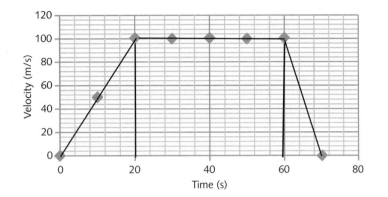

Find the area of each segment and then add them up to get the total change in displacement.

1. Triangle from $t = 0$ s to $t = 20$ s: $\text{area} = \frac{1}{2}(100 \text{ m/s})(20 \text{ s}) = 1000 \text{ m}$

2. Rectangle from $t = 20$ s to $t = 60$ s: $\text{area} = (100 \text{ m/s})(40 \text{ s}) = 4000 \text{ m}$

3. Triangle from $t = 60$ s to $t = 70$ s: $\text{area} = \frac{1}{2}(100 \text{ m/s})(10 \text{ s}) = 500 \text{ m}$

Total displacement = +5500 m

Important Tip

If the velocity vs. time graph includes *negative* velocities, then the portion below the horizontal axis would be evaluated as a *negative* area and correspond to a *negative* change in displacement (backward motion).

Problem Using the velocity vs. time graph shown, determine the final displacement of the object.

Solution

Step 1. The goal of this problem is to determine the final displacement of the object. We will use the method of areas again because this is a velocity vs. time graph. However, in this case, we identify the fact that some of the velocities are *negative* (indicating backward motion).

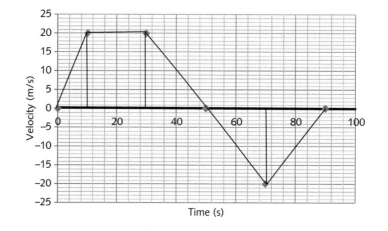

Time (s)

Step 2. We will break the entire region up into four subregions and find the areas of those.

a. From $t = 0$ s to $t = 10$ s: area of a triangle $= \frac{1}{2}bh = \frac{1}{2}(10$ s$)(20$ m/s$) = 100$ m

b. From $t = 10$ s to $t = 30$ s: area of a rectangle $= bh = (20$ s$)(20$ m/s$) = 400$ m

c. From $t = 30$ s to $t = 50$ s: area of a triangle $= \frac{1}{2}bh = \frac{1}{2}(20$ s$)(20$ m/s$) = 200$ m

d. From $t = 50$ s to $t = 90$ s: area of a triangle $= \frac{1}{2}bh = \frac{1}{2}(40$ s$)(-20$ m/s$) = -400$ m

Step 3. *Add* up the displacements (recognizing that the last area is negative).

100 m + 400 m + 200 m + (–400 m) = +300 m is the final displacement.

Problem-Solving Strategies to Avoid Missteps

When solving motion problems it is very important to remember code words such *dropped, falling, rest,* or *stop,* which imply certain values for kinematic quantities. It is also important to remember to include units in all calculations and final answers. The differences between vector and scalar quantities are also important because direction changes imply changes in sign. As in all problems, refer back to the problem-solving ring, and make sure you understand the definitions of key terms and concepts. Identify your given information, identify your goal, understand what the question is asking, and apply your conceptual and mathematical knowledge to solve the problem. Exercise 4.1 ranges in level of difficulty and allows you to test your understanding.

Exercise 4.1

1. A person walks 10 km west during a time interval of 2.5 hours. What is the person's average velocity in m/s?

2. A car is going at 25 m/s and accelerates uniformly to a velocity of 60 m/s in 7 s.

   a. Calculate the acceleration of the car.

   b. Calculate the distance traveled by the car during that time.

3. An object is dropped from the top of a 100-m-tall cliff.

   a. How long will it take the object to reach the ground?

   b. What will be the velocity of the object as it hits the ground?

4. An object is thrown upwards with a velocity of +30 m/s.

   a. What is its displacement after 2.0 s?

   b. How fast will it be going at that time?

   c. How high will it go?

5. Given the velocity vs. time graph shown:

   a. How far will the object have traveled during the 60-s time interval?

   b. Calculate the acceleration of the object.

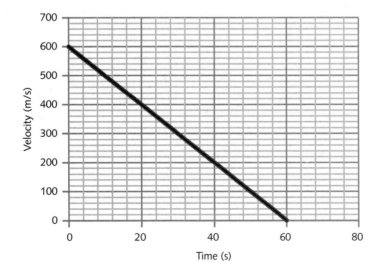

6. Data for the position of an object is shown.

TIME (S)	POSITION (M)
0	0
4	12
8	16
12	18
16	20
18	10
20	0

a. On the grid shown, make a graph of displacement vs. time.

b. What was the average velocity of the object from $t = 0$ s to $t = 4$ s?

c. What was the total distance traveled by the object during the entire 20-s interval?

d. What was the average *speed* of the object during that total time?

e. What was the average *velocity* of the object during that total time?

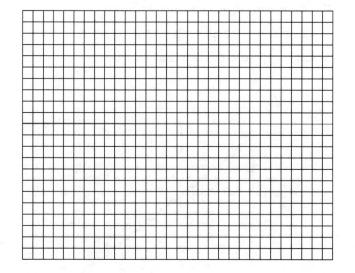

5

Vectors

In this chapter you will learn about how vector quantities are analyzed using geometry and trigonometry. This is a very important step in our journey of exploration of the mechanical universe.

Review of Trigonometry

You should recall that a *triangle* (Figure 5.1) is any closed polygon with three sides. The sum of the angles in any triangle will be equal to 180°. A triangle with two equal sides is called an *isosceles* triangle. In an isosceles triangle, the angles opposite to the equal sides are also equal. A triangle that contains a 90° angle (the word *perpendicular* is also used) is called a *right* triangle (the two other angles are both equal to 45°). An *equilateral* triangle is a triangle with three equal sides (each angle is equal to 60°). Angles are usually designated by capital letters, while corresponding sides are designated by lowercase letters.

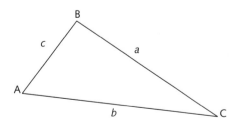

Figure 5.1 A triangle

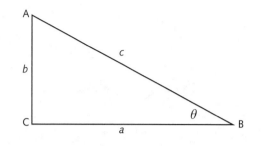

Figure 5.2 Right triangle

The Right Triangle

In a right triangle (see Figure 5.2) the two sides and the hypotenuse are related by the *Pythagorean Theorem*:

$$c^2 = a^2 + b^2$$

The angles and the sides are also related by the following definitions (for angle θ):

sine $\theta = b/c$ cosine $\theta = c/a$ tangent $\theta = b/a$

These relationships are often associated with the mnemonic SOHCAHTOA, which means

Sine = opposite/hypotenuse   Cosine = adjacent/hypotenuse
Tangent = opposite/adjacent

Problem   Find the missing sides in the following right triangles (no units).

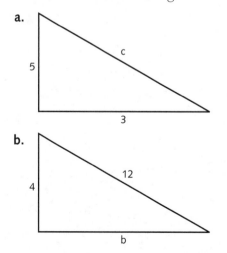

Solution

a.

Step 1. The goal of the problem is to solve for the missing side using the Pythagorean Theorem. Substitute in the given values:

$$c^2 = a^2 + b^2 = 5^2 + 3^2$$

Step 2. Solve for the missing side:

$$c^2 = 5^2 + 3^2 = 25 + 9 = 34$$

$$c = \sqrt{34} = 5.83$$

b.

Step 1. The goal of the problem is to solve for the missing side using the Pythagorean Theorem. Substitute in the given values:

$$c^2 = a^2 + b^2$$

$$12^2 = 4^2 + b^2$$

Step 2. Solve for the missing side:

$$144 = 16 + b^2$$

$$b^2 = 128$$

$$b = 11.31$$

Problem Given the following right triangle (no units), find:

a. The value of angle θ

b. The value of side b using the tangent function

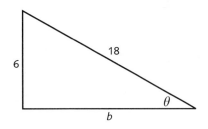

Solution

a.

Step 1. The goal of this problem is to find the angle θ. Given the opposite side and the hypotenuse, we can use the sine function.

Step 2. Insert the given values and solve for θ. You will have to find the inverse sine, or the arcsine, since you will be given the value of the sine of an unknown angle. Use your calculator for this:

$\sin \theta = 6/18 = 0.3333$

$\theta = \sin^{-1}(0.3333) = 19.47°$

b.

Step 1. We need to use the value of the angle obtained in part (a) and use the tangent function to get side b (instead of using the Pythagorean Theorem):

$\tan \theta = $ opposite/adjacent

Step 2. Substitute the known values and solve for side b:

$\tan(19.47°) = 6/b$

and, therefore,

$b = 16.97$

Misconception

When using a scientific calculator, make sure you are aware of how to enter the proper keystrokes. Finding the values of trigonometric and inverse trigonometric functions is different on each calculator. It is also important to remember that in an introductory physics class, the angles are measured in degrees and not radians (which are angle measurements related to a unit circle). Finally, the concept of SOHCAHTOA applies only to *right* triangles.

Non-Right Triangles

If a given triangle does not contain a 90° angle (Figure 5.3), then the sides and angles are related through rules known as the *Law of Cosines* and the *Law of Sines*:

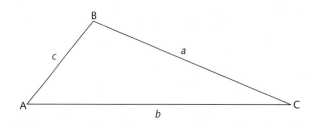

Figure 5.3 Non-right triangle

Law of Cosines

$$c^2 = a^2 + b^2 - 2ab\,\cos(C)$$

Law of Sines

$$\frac{a}{\sin A} = \frac{b}{\sin B} = \frac{c}{\sin C}$$

Problem   For the triangle shown, find the values of side c and angle A using the Law of Cosines and the Law of Sines.

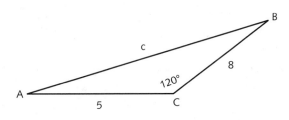

Solution

Step 1. The goal of this problem is to solve for side c and angle A. Start with the law of cosines:

$$c^2 = a^2 + b^2 - 2ab\,\cos(C)$$

Step 2. Substitute in the given values and solve for side c:

$$c^2 = 5^2 + 8^2 - 2(5)(8)\cos(120°)$$

$$c^2 = 25 + 64 - (80)(-0.5) = 25 + 64 + 40 = 129$$

$$c = \sqrt{129} = 11.36$$

Step 3. Use the value for side c to find the value of angle A (use the Law of Sines):

$$\frac{8}{\sin A} = \frac{11.36}{\sin(120)}$$

$$\sin A = (8)\sin(120)/11.36 = (8)(0.8660)/(11.36) = 0.6099$$

$$A = \sin^{-1}(0.6099) = 37.58°$$

Misconception

It is important to watch the sign changes for the values of sines and cosines in the different quadrants (for angles greater than 90°). Notice that in the earlier problem, the value of $\cos(120°) = -0.5$, which affected the negative sign in the Law of Cosines formula.

Combining Vectors

In the last chapter we defined vectors as quantities that have *both* magnitude (size) and direction. Some examples of vectors were displacement, velocity, acceleration, and force. We also saw how these quantities differed from typical scalars (which only have magnitude) like distance, speed, mass, and area.

Because vectors have direction, combining them (known as *vector addition*) is not the same as the addition of scalars (adding ordinary numbers). The result of a combination of vectors is called the *resultant*. We can use our knowledge of kinematics to help explain these ideas.

Recall that vectors are geometrically represented by directed line segments. The length of the line segment (which can be scaled to fit on your paper) represents the magnitude (or scalar portion) of the vector. The arrowhead indicates the direction in the chosen frame of reference (coordinate system). The arrowhead of the vector is simply referred to as the *head*; the end of the vector is referred to as the *tail*.

Problem A person walks 3 m east and then 5 m east. What is the final displacement of the person?

Solution

Step 1. The goal is to find the vector quantity known as *displacement* (see Chapter 4). This is a vector quantity and the situation could be sketched (or drawn to scale, with 1 cm = 1 m) as shown (the sketch is *not* drawn to actual scale).

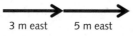

3 m east   5 m east

Step 2. Because the two vectors are sequential and in the same direction, the resultant is just equal to the sum of the two magnitudes and the direction is the same as the original vectors: **d** = resultant displacement = 8 m east.

Important Tip

We can generalize the problem to state that the magnitude of the resultant of any two vectors that are acting in the *same* direction is equal to the *sum* of their individual magnitudes. It is also important to recognize that when combining vectors geometrically, they must be connected *head to tail*.

Problem   A person walks 3 m east and then walks 5 m west. What is the final displacement of the person?

Solution

Step 1. The goal is to find the vector displacement. This is a vector quantity and the situation could be sketched (or drawn to scale, with 1 cm = 1 m) as shown (the sketch is *not* drawn to actual scale).

d = resultant displacement = 2 m west

Step 2. From the sketch, we can see that the final resultant of the combination of the two vector displacements is equal to 2 m west (where we subtracted the opposite magnitudes). The final direction is *west* because the larger of the two vectors is in that direction.

Important Tip

We can again generalize the example to state that if two vectors act in the *opposite* direction, then the magnitude of their resultant is equal to the *difference* between their individual magnitudes. We can also state that the resultant of two vectors is a maximum when they are acting in the same direction and is a minimum when they are acting in the opposite direction. If two vectors are equal in magnitude, but opposite in direction, then their resultant is equal to zero (see Chapter 6 on forces and equilibrium).

Misconception

It is important to remind you about the differences between *distance* and *displacement* as discussed in Chapter 4. In the first example, the *distance* traveled by the person is 8 m (which is the same magnitude of the final *displacement*). In the second example, the *distance* traveled by the person was again equal to 8 m (but the magnitude of the final *displacement* was only 2 m!).

Problem   A person walks 3 m east and then 5 m north. What is final displacement of the person?

Solution

Step 1. The goal of this problem is to find the vector resultant that corresponds to the final displacement of the person. We will again use a

sketch and solve the problem using trigonometry (you could use a scaled drawing with a ruler and a protractor). This time, we will need to find the angle θ, which represents the direction of the resultant.

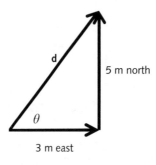

5 m north

3 m east

Step 2. We can see that we have a right triangle. The magnitude of the final displacement will be found using the Pythagorean Theorem. Substitute in the given values and units:

$$\mathbf{d}^2 = (3 \text{ m})^2 + (5 \text{ m})^2 = 34 \text{ m}^2$$

$$\mathbf{d} = 5.83 \text{ m}$$

Step 3. To find the angle θ, the direction of the resultant (note that the displacement vector is drawn from the starting point to the ending point). We can use the tangent function:

$$\tan \theta = 5 \text{ m}/3 \text{ m} = 1.67, \quad \theta = 59.1° \text{ northeast}$$

Problem A person walks 3 m east and then 5 m at 30° northeast. What is the final displacement of the person?

Solution

Step 1. The goal of this problem is to find the resultant displacement. In this case we do not have a perpendicular set of displacements. A sketch of the situation is shown.

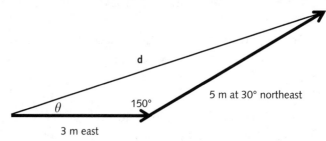

d

150°

5 m at 30° northeast

θ

3 m east

Step 2. The sketch shows a non-right triangle with 150° as the interior angle. Use the Law of Cosines to find the magnitude of the final displacement. Substitute in the given values and units:

$$\mathbf{d}^2 = (3 \text{ m})^2 + (5 \text{ m})^2 - 2(3 \text{ m})(5 \text{ m})\cos(150°) = 59.98 \text{ m}^2$$

$$\mathbf{d} = 7.74 \text{ m}$$

Step 3. To find the direction angle for the resultant displacement, find the angle between the 3-meter displacement and the resultant using the Law of Sines (and the value of the resultant).

$$\frac{5 \text{ m}}{\sin \theta} = \frac{7.74 \text{ m}}{\sin(150)}$$

$$\sin \theta = 0.323 \quad \text{and} \quad \theta = 18.84°$$

Important Tip

If two vectors act on the same point and at the same time, then they are called *concurrent* vectors. To solve problems with concurrent vectors, we use the *parallelogram* method. If you were making a scaled drawing, then a parallelogram (instead of a triangle) would have to be constructed. Velocity and force vectors are typically *concurrent* (whereas displacement vectors are *sequential*).

Misconception

It will be important to recall that for vector addition the vectors need to be connected *head to tail*. This ensures that the correct diagonal of the parallelogram is analyzed so as to obtain the correct direction of the resultant.

Problem   A boat is heading north at 10 m/s crossing a river that is flowing east at 5 m/s. What is the resultant velocity of the boat (as seen by somebody standing on the shore; this is also known as the *relative* velocity)? A sketch of the problem is shown.

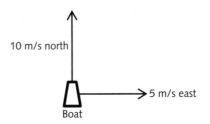

Solution

Step 1. The goal of this problem is to find the resultant velocity. The two velocity vectors are concurrent, so the sketch needs to be redrawn to make sure that the vectors are *head to tail*. We can then use the Pythagorean Theorem, since the vectors are perpendicular to each other.

Step 2. Redraw the sketch as a parallelogram. If you wanted to make a scaled drawing, you could use a scale of 1 cm = 1 m/s. We will just make a sketch and solve using trigonometry.

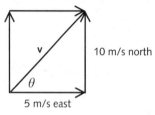

Step 3. Use the Pythagorean Theorem to solve for the resultant velocity:

$$\mathbf{v}^2 = (5 \text{ m/s})^2 + (10 \text{ m/s})^2 = 125 \text{ m}^2/\text{s}^2$$

$$\mathbf{v} = 11.18 \text{ m/s}$$

Step 4. The direction angle, θ, can be found using the tangent function.

$$\tan \theta = (10 \text{ m/s})/(5 \text{ m/s}) = 2$$

$$\theta = 63.43°$$

Components of Vectors

Any vector can be resolved (broken up) into two perpendicular components (parts). The horizontal and vertical components can be found using trigonometry and will be important when we discuss forces and projectiles in later chapters (see Figure 5.4).

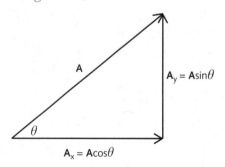

Figure 5.4 Diagram of vector components

Problem A golf ball is hit with a velocity of 80 m/s at an angle of 40° to the horizontal (as sketched). What are the magnitudes of the horizontal and vertical components of the velocity?

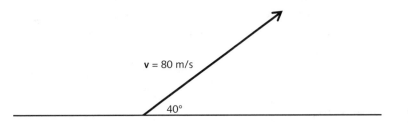

Solution

Step 1. We can redraw the sketch in the form of a right triangle. The goal of the problem is to find the two components. These are basically the sides of a right triangle whose angle and hypotenuse are given. We can call the horizontal component $\mathbf{v}_x$ and the vertical component $\mathbf{v}_y$:

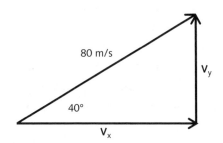

Step 2. We can use SOHCAHTOA to solve for the two components.

$$\cos(40°) = \mathbf{v}_x/80 \text{ m/s}$$

$$\mathbf{v}_x = (80 \text{ m/s})\cos(40°) = 61.28 \text{ m/s}$$

$$\sin(40°) = \mathbf{v}_y/80 \text{ m/s}$$

$$\mathbf{v}_y = (80 \text{ m/s})\sin(40°) = 51.44 \text{ m/s}$$

Because vectors acting in the same direction simply add up, more than two vectors can be combined using a component method. That is, the resultant of three or more vectors can be found by simply adding up the appropriate horizontal and vertical components independently.

Problem-Solving Strategies to Avoid Missteps

When solving vector problems, it is important to remember that when making a scaled drawing (or a sketch), the vectors must be connected *head to tail*. Use the Pythagorean Theorem and SOHCAHTOA for right triangles and the Laws of Sines and Cosines for non-right triangles. You should refer to the problem-solving ring to define goals, as needed, and make sure to include all units when making substitutions.

Exercise 5.1

1. A person walks 9 km east and then 14 km at 30° northwest. What is the magnitude and direction of final displacement of the person (solve by trigonometry).

2. A plane is flying at a velocity of 250 m/s at an angle of 53° southeast. What are the values of the south and east components of the velocity?

3. A projectile is fired into the air at an unknown angle. It is observed that it has a horizontal launch velocity of 25 m/s and a vertical launch velocity of 35 m/s. Determine the value of the launch velocity and the value of the launch angle.

4. A baseball is hit at a 53° angle to the horizontal with a velocity of 78 m/s. Calculate the vertical and horizontal components of the baseball.

5. A person walks 4 km north, 5 km east, and then 8 km west. What is the magnitude and direction of the final displacement of the person?

6

Forces

In this chapter you will learn about forces and the concept of dynamics. Forces interact with matter according to *Newton's Three Laws of Motion*. This is an important development in what we have learned so far. Kinematics is a descriptive study of how objects move. In dynamics, we study the way forces change the motion of objects. It is the next step in our journey.

Types of Forces

In 1687, Isaac Newton (1642–1727) published his *Principia Scientifica* (or *The Mathematical Principles of Natural Philosophy*). In that book, Newton established the concepts of the subject we now know as *classical mechanics* (or *dynamics*). Galileo's contributions to the study of kinematics (how objects move) was now advanced by Newton to include the concept of force and why objects change their motion.

Forces are vector quantities (pushes or pulls) that result when bodies interact (such as in a collision). This means we can use the rules of vector analysis (studied in Chapter 5) to explore how forces change the motion of objects. Later on, we will learn that forces are the agents by which energy is transferred (through the process of work).

In modern times, physicists have identified four *fundamental forces*:

1. Gravitational force

2. Electromagnetic force

3. Strong nuclear force

4. Weak nuclear force

Gravitation is the name given by Newton to the force that pulls objects toward the earth. He realized that the force that keeps the moon in orbit around the earth is the same force acting on any falling object. The *electromagnetic force* (unified by James Clerk Maxwell in the late nineteenth century) is the name of the force that links electricity and magnetism. Electromagnetism plays an important role in the study of optics, and most of our common interactions are actually electromagnetic in nature. Because like charges repel (see Chapter 16), the inability of one object to pass through another object is due to the electrical repulsion of electrons in matter.

The last two fundamental forces were discovered within the past 70 years. The *strong nuclear force* is the name given to the force that binds protons and neutrons together inside an atomic nucleus. The *weak nuclear force* is the name given to the force that governs radioactive decay. We will return to these concepts in Chapter 20.

Recently, physicists have been pursuing a unification of forces in an attempt to better understand the fundamental nature of the universe. In 1979, the Nobel Prize in Physics was awarded to Abdus Salam, Steven Weinberg, and Sheldon Glashow for theoretically unifying the electromagnetic and weak forces into what is now called the *electro-weak force* at high energies.

Newton's Laws

In his *Principia*, Isaac Newton expressed his ideas about forces and motion in his famous three laws:

1. Every object perseveres in its state of rest, or of uniform motion in a straight line, unless it is compelled to change that state by forces impressed upon it.

2. The alteration of motion is proportional to the motive force impressed and is made in the direction of the straight line in which that force is impressed.

3. To every action there is an equal and opposite reaction. Or, the mutual actions of two objects upon each other are always equal and directed to contrary parts.

The First Law

The first law describes the concept now known as inertia. *Inertia* is defined as the tendency of an object to resist an unbalanced (or net) force acting on it. It is important to remember that this and the other laws are based on a given frame of reference.

The first law (also known as the *Law of Inertia*) can be rephrased in a more common form: "An object at rest tends to stay at rest, unless acted upon by an external net force. An object in motion tends to stay in motion, at constant velocity, unless acted upon by an external net force."

Based on what we know about vectors, the concept of net force can be related to the concept of *vector resultant*. The *net force* is simply the vector sum of all forces acting on an object. On the other hand, inertia is a *scalar* quantity that is often associated with the mass (also a scalar) of the object (objects that have more mass have more inertia and tend to resist net forces changing their state of motion).

Examples of the First Law

1. If you are sitting in a car at rest, you feel as though you are being pushed backward when the car accelerates forward.

2. If you are in a car that suddenly stops, you feel as though you are being pushed forward.

3. If the car you are traveling in makes a sharp right turn, you feel as though you are being pushed to the left side of the car.

We can now understand why an object with zero acceleration may or may not be at rest (in a given frame of reference). According to the Law of Inertia, an object that does not have a net force acting on it maintains constant velocity (zero acceleration).

 Misconception

Inertia is *not* a force and it is *not equal* to the mass of the object. Also, force and mass are *not* the same thing. There is no force pushing you back into the seat of the car as it accelerates forward. It is your belief in the concept of inertia that allows you to develop this idea of a *fictitious* force. The absence of a net force does not mean that are no forces acting on an object. It only means that the *net* (vector sum) resultant of all forces is equal to zero. Finally, once the net force stops acting, the object will continue to move, but only in a straight line at a constant velocity (see Chapter 7).

The Second Law

The second law is more commonly known as the *Law of Acceleration*:

$$\Sigma \mathbf{F} = \mathbf{F}_{net} = m\mathbf{a} \text{ (the symbol } \Sigma \text{, means "sum" and "m" represents the mass of the object)}$$

The unit of force is called a Newton (N) and is defined by the second law:

$$1 \text{ N} = 1 \text{ kg} \cdot \text{m/s}^2 \text{ (units of mass must be in kilograms)}$$

Some examples of the second law would be observed when you push an object: the greater the force, the greater the acceleration. An object sliding along a rough floor is slowed by the force of "friction." The moon orbiting around the earth maintains an elliptical path due to the force of gravity changing its direction and velocity.

The action of a *net force* is to change the velocity of an object (that is, it can change its speed and/or its direction). The direction of the acceleration is the *same* as the direction of the net force acting on the mass. If an object is sliding on top of a surface, then *friction* is the name of the force that opposes the motion of the object, as it slides over the surface.

Newton's second law can also be written in the form:

$$\mathbf{F} = m\left(\frac{\Delta \mathbf{v}}{\Delta t}\right)$$

or

$$\mathbf{F}\Delta t = m\Delta \mathbf{v}$$

This equation will be important when we study the concepts of projectile motion, circular motion, and momentum in Chapters 7 and 9.

If the net force is due to gravity, and the object is falling near the surface of the earth (recall $\mathbf{a} = \mathbf{g} = -9.8 \text{ m/s}^2$), then we can write:

$$\mathbf{F}_g = m\mathbf{g}$$

This force is also identified as the *weight* (which is also a force and a vector) of the object when it is standing on the earth.

Problem A 5-kg mass is shown with two forces acting on it. It has an initial velocity of 10 m/s to the right. Determine the magnitude and direction of both the net force and acceleration acting on the mass, as well as the velocity of the object after the net force has acted for 5 seconds.

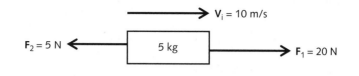

Solution

Step 1. To find the net force, you need to recognize that what the problem is really asking for is the vector sum of the two forces, which (in this case) are acting in opposite directions. Using our rules for vector addition (use + to represent the *right* direction), the net force is simply equal to:

$$\Sigma F = \text{sum of all forces} = F_{net} = 20\ N + (-5\ N) = 15\ N\ (\text{right})$$

Step 2. To find the acceleration, use the second law:

$$F_{net} = 15\ N = (5\ kg)a$$

and

$$a = 3\ m/s^2\ (\text{right})$$

Step 3. To find the new velocity, recall that $\Delta v = v_f - v_i$ and use the acceleration equation:

$$a = \left(\frac{v_f - v_i}{t}\right) \text{ or } v_f = v_i + at$$

$$v_f = (10\ m/s) + (3\ m/s^2)(5\ s) = 25\ m/s\ (\text{right})$$

Important Tip

We can now better understand why an object slowing down (in the forward + direction) has negative acceleration. If an object is moving forward and it is subject to an opposing net force (like friction), then the acceleration of the object will be negative.

Problem   What is the weight of a 5-kg mass as it sits on a table?

Solution

Step 1. The goal of the problem is to find the value of the gravitational force acting on the mass:

$$\text{Weight} = F_g = mg$$

Step 2. Substitute in the given values with units:

$$F_g = (5\ kg)(-9.8\ m/s^2) = -49\ N\ (\text{directed downward})$$

The Third Law

The third law is commonly known as the *Law of Action and Reaction.*

Examples of the Third Law

1. The air in an expanded balloon is released and the balloon goes in the opposite direction.

2. When rowing a boat, the water is pushed backward and the boat goes forward.

3. When launching a rocket, the ignited fuel is ejected toward the ground and the rocket is thrust upward.

Problem   A rocket has a mass of 500 kg. It is launched with an upward acceleration of +20 m/s². What is the magnitude of the upward thrust acting on the rocket?

Solution

Step 1. The goal of this problem is to find the magnitude of the rocket's thrust. However, in order to do this, we must first determine the weight of the rocket, which acts against the thrust.

$$\mathbf{F}_g = m\mathbf{g} = (500 \text{ kg})(-9.8 \text{ m/s}^2) = -4900 \text{ N}$$

Step 2. The net force acting on a mass is equal to the product of the mass × acceleration:

$$\mathbf{F}_{net} = m\mathbf{a} = (500 \text{ kg})(+20 \text{ m/s}^2) = +10,000 \text{ N}$$

Step 3. The definition of the net force is that it is equal to the sum of all the vector forces acting on the mass:

$$\mathbf{F}_{net} = \Sigma\mathbf{F} = \mathbf{F}_{Thrust} + \mathbf{F}_g = \mathbf{F}_{Thrust} + (-4900 \text{ N}) = 10,000 \text{ N}$$
$$\mathbf{F}_{Thrust} = 14,900 \text{ N}$$

Equilibrium

If the net force acting on a mass is equal to zero, then the object is said to be in *equilibrium.* If the object is at rest, then the object is in *static equilibrium.* The object shown in Figure 6.1 is in static equilibrium. The support force of the table (acting upward) balances the force of gravity (acting downward). This upward force (acting perpendicular to the surface) is also known as the *normal force.*

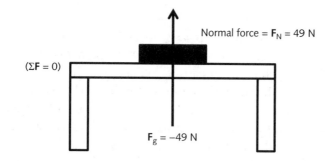

Figure 6.1 Static equilibrium for a mass on a table

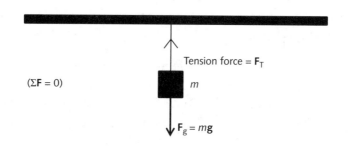

Figure 6.2 Static equilibrium for a suspended mass from a string

An object that is moving with a constant velocity is in a state of dynamic equilibrium. If the object is not in equilibrium, then the application of a force equal and opposite to the net force can place the object into a state of equilibrium. This force is known as the *equilibrant force*. A suspended object is also in equilibrium. In this case, the upward force is provided by the tension in the string (see Figure 6.2).

Problem A 5-kg mass is being pulled along the ground by a string that makes a 30° angle to the horizontal. Assuming negligible friction between the mass and the ground, calculate the magnitude of the horizontal acceleration of the mass if the applied force in the string is equal to 80 N.

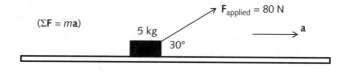

Solution

Step 1. In this problem, there are two goals. In one case, the object is not moving vertically, which means that the net vertical force is equal to zero. In the second case, the object will have a horizontal acceleration (as stated) and, therefore, the net horizontal force needs to be determined. However, the only force explicitly given is the tension in the string (which is acting at an angle). To solve all of these issues, we need to identify the sum of all vector forces both horizontally *and* vertically:

$$\Sigma \mathbf{F}_x = m\mathbf{a}_x$$
$$\Sigma \mathbf{F}_y = 0$$

Step 2. Identify all of the forces acting horizontally and vertically using vector components:

In the horizontal (x) direction: $\mathbf{F}_x = \mathbf{F}\cos\theta = (80\text{ N})\cos(30°) = 69.28\text{ N}$ (right)

In the vertical (y) direction: $\mathbf{F}_g = m\mathbf{g} = (5\text{ kg})(-9.8\text{ m/s}^2) = -49\text{ N}$ (downward)

$\mathbf{F}_y = \mathbf{F}\sin\theta = (80\text{ N})\sin(30°) = 40\text{ N}$ (upward)

Step 3. Since the upward vertical force component of the string is less than the downward force of gravity, the object will not leave the ground (no vertical acceleration). The only horizontal force acting on the mass is the horizontal component of the applied force ($\mathbf{F}_x$). Therefore,

$$\mathbf{F}_x = m\mathbf{a}_x$$
$$69.28\text{ N} = (5\text{ kg})\mathbf{a}_x$$
$$\mathbf{a}_x = 13.86\text{ m/s}^2$$

Problem-Solving Strategies to Avoid Missteps

Sometimes, it is useful to identify all of the forces acting on a mass using a free-body diagram. In this type of analysis, you draw a separate diagram (see Figure 6.3) and isolate the mass. Draw all of the given forces (*not* the components), and then begin your problem solving using Newton's second law. In the earlier example, since there is no vertical acceleration, the normal force ($\mathbf{F}_N$) would be equal to (and act as an equilibrant force) the difference between the downward force of gravity and the upward vertical component of the tension in the string:

$$\Sigma \mathbf{F}_y = \mathbf{F}_N + \mathbf{F}_g + \mathbf{F}_y = \mathbf{F}_N + (-49\text{ N}) + 40\text{ N} = 0$$

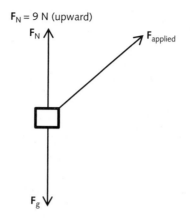

$F_N = 9$ N (upward)

Figure 6.3 A free-body diagram

Hooke's Law

Forces are measured using spring scales. These are calibrated in units of Newtons because of a relationship discovered by Robert Hooke (a scientist who lived at the same time as Isaac Newton). Imagine a spring suspended vertically. As seen in Figure 6.4, if a mass is attached to the spring, it will stretch (elongate) by an amount Δx. The force applied is equal to the weight of the mass ($\mathbf{F}_g = m\mathbf{g}$). The spring also exerts a restoring force upward ($\mathbf{F}_s$).

The ability of the spring to return to its original length, after the force is removed, is an indication of its *elasticity*. Hooke was able to show that there was a direct relationship between the applied force and the elongation (amount of stretch) of the spring, as shown in Figure 6.5.

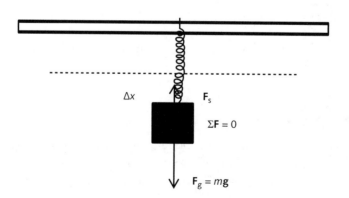

Figure 6.4 Hooke's Law for stretching a spring

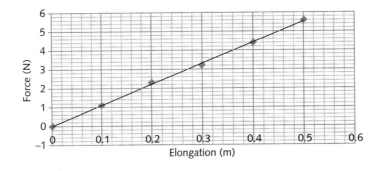

Figure 6.5 Force vs. elongation for a stretched spring

In this example, the equation for Hooke's Law is given by:

$$\mathbf{F} = k\Delta x$$

The constant k is called the *spring constant* (or the *force constant*), in units of N/m. For Figure 6.5, the spring constant is found by taking the slope of the line (approximately):

$$k = \Delta \mathbf{F}/\Delta x = 11 \text{ N/m}$$

If this spring now is subjected to a 0.5-N force, then the spring will stretch by an amount

$$0.5 \text{ N} = (11 \text{ N/m})\Delta x$$

and, therefore,

$$\Delta x = 0.05 \text{ m}$$

Forces on Inclines

If a mass is sliding down an incline, then the forces acting on the object need to be analyzed differently since the frame of reference has changed. As shown in Figure 6.6, the force parallel to the incline is given by

$$\mathbf{F}\| = m\mathbf{g} \sin \theta$$

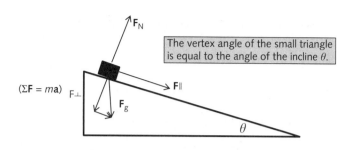

Figure 6.6 Forces acting on a mass sliding down an inclined plane without friction

and the force acting perpendicular to the incline is given by

$$\mathbf{F}\| = mg \cos \theta$$

In this case, the normal force would be equal (and opposite to) the perpendicular force, $\mathbf{F}\|$.

Problem A 5-kg mass, initially at rest, slides down a frictionless 30° incline. Calculate the force parallel to the incline and the acceleration of the mass. If the incline is 0.8 m long, calculate the velocity of the mass when it reaches the bottom of the incline.

Solution

Step 1. The goal of the problem is to calculate the desired quantities of force, acceleration, and velocity parallel to the incline (when friction is neglected).

If the object were resting on the ground, then gravity would be able to accelerate it forward. However, because it is on an incline, gravity now has a component force parallel to the incline which can accelerate the mass. You must therefore change the frame of reference to solve this problem and use the appropriate equations for an accelerating mass on an incline.

Step 2. We can substitute our given values (with units) into the equation.

$$\mathbf{F}\| = mg \sin \theta = (5 \text{ kg})(-9.8 \text{ m/s}^2)\sin(30°) = -24.5 \text{ N (down the incline)}$$

Step 3. Newton's second law is adapted for the motion along an incline. In this case:

$$\mathbf{F}_{net} = m\mathbf{a}$$

and, therefore,

$$\mathbf{F}_{net} = \mathbf{F}\| = 24.5 \text{ N} = (5 \text{ kg})\mathbf{a}\|$$
$$\mathbf{a}\| = -4.9 \text{ m/s}^2 \text{ (down the incline)}$$

Step 4. We can use the equation (knowing that $\mathbf{v}_i = 0$) $\mathbf{v}_f^2 - \mathbf{v}_i^2 = 2\mathbf{a}d$, where the acceleration is parallel to the incline.

$$\mathbf{v}_f^2 = 2(-4.9 \text{ m/s}^2)(0.8 \text{ m})$$
$$\mathbf{v}_f = -2.8 \text{ m/s (down the incline)}$$

Friction

Friction is a force that opposes motion as an object slides over a surface. When an object is at rest, there is a minimum force needed to get the object started. If the object is lying flat on the surface, then the normal force is just equal in magnitude to the weight of the object. The minimum force needed to get the object to start moving is a measure of the *maximum static friction* acting between the two surfaces (see Figure 6.7).

There is a direct relationship between the normal force and the maximum static friction between two surfaces:

$$\mathbf{f}_{s,max} = \mu_s \mathbf{F}_N$$

where μ_s is called the coefficient of static friction. Static friction is independent of surface areas in contact.

When an object is sliding over a surface, then *kinetic friction* is the force that opposes the applied force. The kinetic friction can be measured by finding

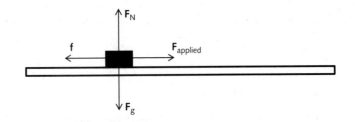

Figure 6.7 Static friction

the force needed to maintain a constant velocity ($\mathbf{F}_{net} = 0$). The diagram in Figure 6.7 can also be used to represent kinetic friction (which is generally less than static friction).

Once again, there is a direct relationship between the kinetic friction and the normal force:

$$\mathbf{f}_k = \mu_k \mathbf{F}_N$$

where μ_k is the coefficient of kinetic friction. Typical coefficients of static and kinetic friction are shown in Table 6.1.

Problem A 5-kg mass is sliding along a frictionless horizontal surface at a velocity of 20 m/s. It suddenly encounters a region with friction and slows to a stop in 45 s. What is the coefficient of kinetic friction between the two surfaces?

Solution

Step 1. The goal of the problem is to find the coefficient of kinetic friction. However, to find μ_k, we first need to identify the forces acting on the mass. Kinetic friction is the only force acting horizontally that slows the object to a stop. The other two forces are the weight of the mass and the normal force as it slides horizontally along the surface.

Table 6.1 Approximate Coefficients of Static and Kinetic Friction

SURFACE MATERIALS	KINETIC	STATIC
Rubber on concrete (dry)	0.68	0.90
Rubber on concrete (wet)	0.58	*
Rubber on asphalt (dry)	0.67	0.85
Rubber on asphalt (wet)	0.53	*
Rubber on ice	0.15	*
Waxed ski on snow	0.05	0.14
Wood on wood	0.30	0.42
Steel on steel	0.57	0.74
Copper on steel	0.36	0.53
Teflon on Teflon	0.04	*

*These surfaces have no coefficient for static friction.

Step 2. The equation for kinetic friction is

$$\mathbf{f}_k = \mu_k \mathbf{F}_N$$

and

$$\mathbf{F}_g = m\mathbf{g} = (5 \text{ kg})(-9.8 \text{ m/s}^2) = -49 \text{ N}$$

Since the mass is sliding horizontally, $\mathbf{F}_N$ = magnitude of $\mathbf{F}_g$ = 49 N.

Step 3. The net force in this case is equal to the force of kinetic friction,

$$\Sigma\mathbf{F} = \mathbf{F}_{net} = m\mathbf{a} = \mathbf{f}_k = \mu_k \mathbf{F}_N$$

and we know that

$$\mathbf{a} = \left(\frac{\mathbf{v}_f - \mathbf{v}_i}{t}\right) = \left(\frac{0 \text{ m/s} - 20 \text{ m/s}}{45 \text{ s}}\right) = -0.44 \text{ m/s}^2$$

Step 4. We solve for the coefficient of kinetic friction:

$$(5 \text{ kg})(-0.44 \text{ m/s}^2) = -\mu_k(49 \text{ N})$$

where we take the friction force as negative since it opposes the motion:

$$\mu_k = 0.045$$

Problem-Solving Strategies to Avoid Missteps

Problems with forces can be difficult and require planning and analysis. It is very important to always identify all of the forces directly acting on masses. Refer to the problem-solving ring to assist you in planning a solution path to a problem. Remember, weight and mass are not the same thing and that the net force is always in the same direction as the acceleration. Make sure to include all units when making calculations and remember that vector methods may need to be used to obtain the correct components. The normal force always acts perpendicular to the surface (and is directed away from it).

Remember that Newton's second law is valid in all dimensions and that forces are vector quantities. This means that you should decide on a positive direction and write down Newton's second law for both vertical and horizontal dimensions (in equilibrium, the net force is equal to zero).

$$\Sigma\mathbf{F}_x = m\mathbf{a}_x$$
$$\Sigma\mathbf{F}_y = m\mathbf{a}_y$$

Exercise 6.1

1. A 100-kg mass is hung from a rod (considered to be of negligible mass) as shown. The system is in static equilibrium. Calculate the tension in the string if the given angle is 30°.

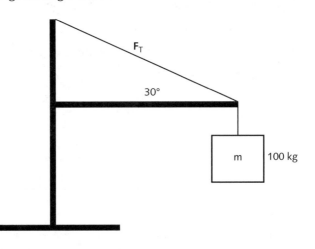

2. A student performs a Hooke's Law experiment and produces the graph shown.

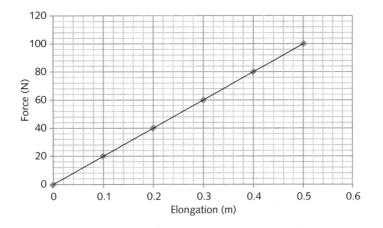

a. Determine the value of the spring constant k.

b. How much force would be required to stretch the spring by 0.27 meter?

3. A 7-kg mass is pulled by a string at an angle of 40° along a surface, as shown. The coefficient of kinetic friction between the mass and the surface is $\mu_k = 0.2$ and the applied force is 50 N.

   a. Determine the weight of the mass.

   b. Determine the horizontal and vertical components of the applied force.

   c. Determine the value of the normal force.

   d. Determine the magnitude of the force of kinetic friction.

   e. Determine the magnitude of the acceleration.

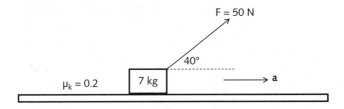

4. A 2-kg mass is sliding down an incline with an unknown angle, as shown. The incline is assumed to be frictionless and is 1.2 m long. The mass starts from rest and is observed to take 2.3 s to reach the bottom. What is the angle of the incline?

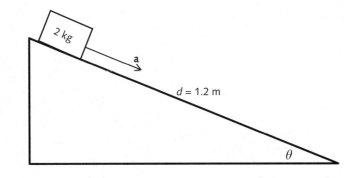

7

Motion in a Plane

In this chapter you will learn about the motion of projectiles and the motion of objects moving in circles. These concepts can then be applied to oscillations and rotational equilibrium. Through applying Newton's Laws of Motion, we will take the next step in our exploration of the mechanical universe.

Horizontal Projectiles

In his book *Dialogues Concerning Two New Sciences*, Galileo Galilei applied his understanding of inertia to the motion of objects in two dimensions. Isaac Newton perfected this analysis in his *Principia*. We can begin our study of projectiles by understanding how a net force changes the path of a projectile into the shape of a parabola. The key to this concept is understanding that vertical and horizontal motions in a plane are independent of each other (following Newton's Laws).

In Figure 7.1 we see an illustration of the phenomenon described by Galileo. A mass (in the absence of any air resistance) is projected horizontally from a given height. If there were no gravity, the motion would be uniformly horizontal (at constant velocity). However, because gravity acts as a force (vertically downward), it creates an acceleration (in the downward direction) and changes the path of the original motion.

The horizontal acceleration is zero (horizontal velocity is constant), so the only acceleration on the mass is $\mathbf{g}$. This means that the horizontal displacement of the projectile is simply:

$$\mathbf{x} = \mathbf{v}_x t$$

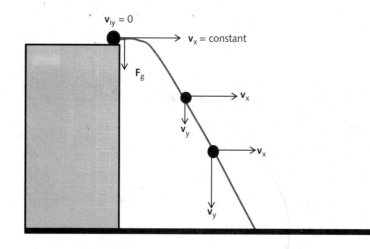

Figure 7.1 Horizontally launched projectile

Because the initial vertical velocity is zero, then the vertical displacement is given by:

$$y = \frac{1}{2}gt^2$$

We can show that this corresponds to the shape of a parabola by expressing the vertical displacement in terms of the horizontal displacement:

$$t = \frac{\mathbf{x}}{\mathbf{v}_x}$$

and, with a substitution,

$$y = \frac{\mathbf{g}\mathbf{x}^2}{2\mathbf{v}_x^2}$$

Because **g** is negative, this is the equation of a concave, downward-pointing semiparabola.

Problem   A projectile is launched horizontally with a velocity of 10 m/s from the roof of a building that is 100 m high. (a) How long will it take the projectile to reach the ground? (b) How far from the base of the building will the projectile land? (Neglect any air resistance.)

Solution

Step 1. We need to recognize that with projectile motion, the two motions are simultaneous and independent. Thus, the goal for part (a) involves only the vertical motion for an object dropped from rest. In this case, $\mathbf{v}_{iy} = 0$, and $\mathbf{a}_y = \mathbf{g} = -9.8$ m/s². We assign the negative to the down direction.

Step 2. We can solve for the time to fall by using the kinematic equation:

$$y = \frac{1}{2}\mathbf{a}t^2 = \frac{1}{2}\mathbf{g}t^2$$

$$-100 \text{ m} = \frac{1}{2}\left(-9.8 \text{ m/s}^2\right)t^2 \text{ (assign the negative to vertical displacement)}$$

$t = 4.52$ s

Step 3. The distance from the base of the roof can now be calculated using the same time found in part (a). The horizontal velocity is constant, and, therefore:

$\mathbf{x} = \mathbf{v}_x t = (10 \text{ m/s})(4.52 \text{ s}) = 45.2$ m

Problem   A ball rolls off of a table that is 0.95 m high. It is observed that it lands 2.5 m away from the base of the table. (a) What was the horizontal velocity as it left the table? (b) What was the value of the vertical velocity as it hit the ground? (Neglect any air resistance.)

Solution

Step 1. The goal of this problem is to find the horizontal velocity, so we need to find the time that the projectile was in the air. During this time, it not only moved forward 2.5 m but also *fell* 0.95 m vertically. Again, there is no horizontal acceleration and the vertical acceleration is given by $\mathbf{g}$. Therefore, we are given that $\mathbf{x} = 2.5$ m, $y = -0.95$ m, and $\mathbf{a}_y = \mathbf{g}$.

Step 2. We solve for the time that it takes the projectile to fall:

$$y = \frac{1}{2}\mathbf{g}t^2$$

$$-0.95 \text{ m} = \frac{1}{2}\left(-9.8 \text{ m/s}^2\right)t^2$$

$t = 0.44$ s

Step 3. Now, we can calculate the horizontal velocity:

$$\mathbf{x} = \mathbf{v}_x t$$

$$2.5 \text{ m} = \mathbf{v}_x (0.44 \text{ s})$$

$$\mathbf{v}_x = 5.68 \text{ m/s}$$

Projectiles Launched at an Angle

We know from our work on kinematics that for an object thrown upward, the time it takes to reach its maximum height (where it momentarily stops) is equal to the time it takes to fall back down to its starting point. For a projectile launched at an angle, the vertical and horizontal motions (neglecting any air resistance) are again simultaneous and independent of each other. In this case, however, there is an initial upward vertical velocity and we must use our techniques of vector analysis if we are given the launch velocity and launch angle. The path of the projectile will be a full inverted parabola (see Figure 7.2). Note that the *vertical* velocity will be equal to *zero* at the maximum height (but $\mathbf{a}_y = \mathbf{g}$ at all times!).

We can calculate the vertical and horizontal displacements using concepts similar to those for horizontal projectiles (recognizing that $\mathbf{v}_{iy}$ is *not* equal to zero):

$$\mathbf{x} = \mathbf{v}_x t$$

$$\mathbf{y} = \mathbf{v}_{iy} t + \frac{1}{2} \mathbf{g} t^2$$

The horizontal and vertical components of the launch velocity can be found using vector analysis. Galileo also demonstrated that the maximum horizontal displacement (called the *range*) of a projectile is itself a maximum when the launch angle is equal to 45°.

The time that it takes the projectile to reach its maximum height is given by the time it takes gravity to decelerate the vertical velocity to zero (recognizing that $\mathbf{v}_y = 0$ at the top):

$$t_{\text{up}} = t_{\text{down}} = -\frac{\mathbf{v}_{iy}}{\mathbf{g}} \quad \text{and} \quad t_{\text{total}} = 2t_{\text{up}}$$

The total time in the air is just equal to twice the time it takes the projectile to reach maximum height.

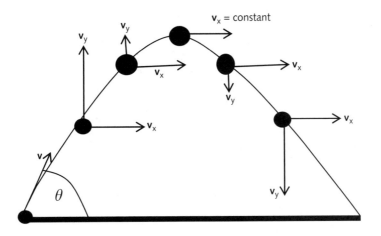

Figure 7.2 Projectile launched at an angle

The maximum height can be found by using t_{up} or by recognizing that because the vertical velocity is equal to zero at the maximum height, the projectile is actually *falling* during its second half (similar to the horizontal projectile problems):

$$\mathbf{y}_{max} = \mathbf{v}_{iy}t_{up} + \frac{1}{2}\mathbf{g}t_{up}^2 \; \left(\text{remember that } \mathbf{g} = -9.8 \text{ m/s}^2\right)$$

$$\mathbf{y}_{max} = \left|\frac{1}{2}\mathbf{g}t_{down}^2\right| \left(\text{since } t_{up} = t_{down}\right)$$

The range will be found using the total time and the horizontal velocity:

$$\text{Range} = \mathbf{x}_{max} = \mathbf{v}_x t_{total} = \mathbf{v}_x\left(2t_{up}\right)$$

Problem A golf ball is launched at a 40° angle with a velocity of 80 m/s. Calculate the horizontal and vertical components of the launch velocity, the maximum height of the golf ball, and the range (maximum horizontal displacement).

Solution

Step 1. In order to perform all of the calculations, we need to evaluate the vertical and horizontal components of the initial launch velocity.

$$\mathbf{v}_{ix} = \mathbf{v}_x = \mathbf{v} \cos \theta = (80 \text{ m/s})\cos(40°) = 61.28 \text{ m/s}$$

$$\mathbf{v}_{iy} = \mathbf{v} \sin \theta = (80 \text{ m/s})\sin(40°) = 51.42 \text{ m/s}$$

Step 2. To find the maximum height, we need to first find the time it takes the projectile to reach the maximum height:

$$t_{up} = -(v_{iy})/g = (51.42 \text{ m/s})/(9.8 \text{ m/s}^2) = 5.25 \text{ s}$$

$$t_{total} = 10.50 \text{ s}$$

$$y_{max} = \left| \frac{1}{2} g t_{down}^2 \right| = \frac{1}{2}(9.8 \text{ m/s}^2)(5.25 \text{ s})^2 = 135.1 \text{ m}$$

Step 3. We use the total time and the horizontal velocity to find the range:

$$\text{Range} = x_{max} = (61.28 \text{ m/s})(10.50 \text{ s}) = 643.44 \text{ m}$$

Problem A football is kicked down the field at an unknown angle. It is observed to land 50 m away after a total of 4.6 s in the air. Calculate the initial horizontal and vertical velocity, the maximum height of the football, and the magnitude of the launch velocity and its launch angle.

Solution

Step 1. In this problem, we are given only the range and the total time. However, we can work backward and determine all of the other information we need.

The horizontal velocity is constant, so $v_{ix} = v_x = (50 \text{ m})/(4.6 \text{ s}) = 10.87 \text{ m/s}$.

Step 2. We know that $t_{total} = 4.6$ s; this means that $t_{up} = 2.3$ s. Because gravity decelerated the initial vertical velocity to zero at the maximum height, it is logical that the magnitude of $v_{iy} = |g t_{up}| = (9.8 \text{ m/s}^2)(2.3 \text{ s}) = 22.54 \text{ m/s}$.

Step 3. To find the maximum height, we can use the time to fall from the maximum height $\left(t_{up} = t_{down} \right): y_{max} = \left| \frac{1}{2} g t_{down}^2 \right| = \frac{1}{2}(9.8 \text{ m/s}^2)(2.3 \text{ s})^2 = 25.92 \text{ m}$.

Step 4. The launch velocity can be found using the Pythagorean Theorem because the vertical and horizontal velocities are perpendicular to each other:

$$v^2 = v_{ix}^2 + v_{iy}^2$$

$$v^2 = (10.87 \text{ m/s})^2 + (22.54 \text{ m/s})^2 = 626.21 \text{ m}^2/\text{s}^2$$

$$v = 25.02 \text{ m/s}$$

The launch angle can be found from the tangent function:

$\tan \theta = (\mathbf{v}_{iy})/(\mathbf{v}_{ix}) = (22.54 \text{ m/s})/(10.87 \text{ m/s}) = 2.074$

$\theta = 64.25°$

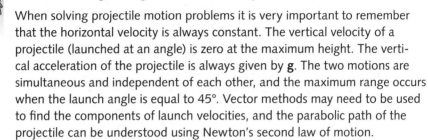

Problem-Solving Strategies to Avoid Missteps

When solving projectile motion problems it is very important to remember that the horizontal velocity is always constant. The vertical velocity of a projectile (launched at an angle) is zero at the maximum height. The vertical acceleration of the projectile is always given by **g**. The two motions are simultaneous and independent of each other, and the maximum range occurs when the launch angle is equal to 45°. Vector methods may need to be used to find the components of launch velocities, and the parabolic path of the projectile can be understood using Newton's second law of motion.

Uniform Circular Motion

As we have seen, a net force can change the path of the motion of an object. If the net force always acts perpendicular to the velocity, the result is a circle. If the speed of the object remains constant, then the motion is called *uniform circular motion*. Remember, even though the speed is constant (the object covers a circumference in the same period of time, T), the object is still said to be accelerating. Isaac Newton recognized this as an application of his Laws of Motion.

Imagine an object connected to string and twirled into a circular path. The tension in the string pulls the object into the path of a circle (see Figure 7.3). This net force is now called the *centripetal force* (centripetal means "center seeking") and is always directed inward toward the center of

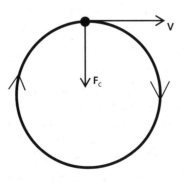

Figure 7.3 Uniform circular motion

the circle. The velocity is tangent to the circle and constant in magnitude (but changing in direction).

If the circle has a radius R, then the centripetal force is given by:

$$\mathbf{F}_c = m\mathbf{a}_c = m\mathbf{v}^2/R$$

where $\mathbf{a}_c = \mathbf{v}^2/R$.

If the period (the time to complete one revolution) of the motion is given by the time T, then the frequency corresponds to the number of revolutions per second and is given by $f = 1/T$.

The centripetal acceleration can also be written in the form:

$$\mathbf{a}_c = 4\pi^2 R/T^2$$

and the centripetal force is now written as:

$$\mathbf{F}_c = m4\pi^2 R/T^2$$

Problem   Calculate the centripetal force for a 0.5-kg mass undergoing uniform circular motion in a path of radius 0.25 m and a period equal to 0.75 s.

Solution

Step 1. The goal of this problem is to find the value of the centripetal force given the mass, radius, and period of the circular motion. We can therefore use the equation

$$\mathbf{F}_c = m4\pi^2 R/T^2$$

Step 2. Substitute the given values with units into the equation:

$$\mathbf{F}_c = (0.5 \text{ kg})4\pi^2(0.25 \text{ m})/(0.75 \text{ s})^2 = 8.77 \text{ N}$$

Misconception

Centripetal force is just another name for a *net force*. In our example, it is the tension in the string that sets up the centripetal force. Friction acts as the centripetal force on a car going around a curve. For an orbiting object, the force of gravity acts as the centripetal force. Additionally, if the net force is removed, then the object will obey Newton's *Law of Inertia*, and follow a straight-line path with constant velocity.

The term *centrifugal force* is sometimes applied to the apparent force (in a rotating frame of reference) that appears to push an object outward and away from the center of the circle. This is a *fictitious* force because the rotating frame of reference is *accelerating* and the *Law of Inertia* is not valid (the same way you experience a force pushing you back into your seat as your car accelerates forward).

Simple Harmonic Motion

Imagine a mass attached to a spring (see Figure 7.4). The mass is resting along a frictionless surface. From Hooke's Law, we know that the spring will provide a restoring force equal to $\mathbf{F}_s = -k\Delta x$ if it is compressed or elongated by an amount Δx.

Because the restoring force of the spring is equal to the net force, we can write:

$$-k\Delta x = m\mathbf{a}$$

and

$$\mathbf{a} = -(k/m)\Delta x$$

The acceleration is equal to the negative of the displacement. When released, the mass will oscillate back and forth through a maximum displacement (called the *amplitude*). This motion is called *simple harmonic motion* and it is similar to what we have learned about circular motion. The period of the oscillation can be derived as follows. If the amplitude of the motion is equal to the radius of an equivalent circular motion (when viewed from the side), then:

$$4\pi^2 R/T^2 = (k/m)R$$

and

$$T_s = 2\pi\sqrt{\frac{m}{k}} \text{ (period varies directly with the square root of mass)}$$

Notice that the period is independent of the acceleration due to gravity.

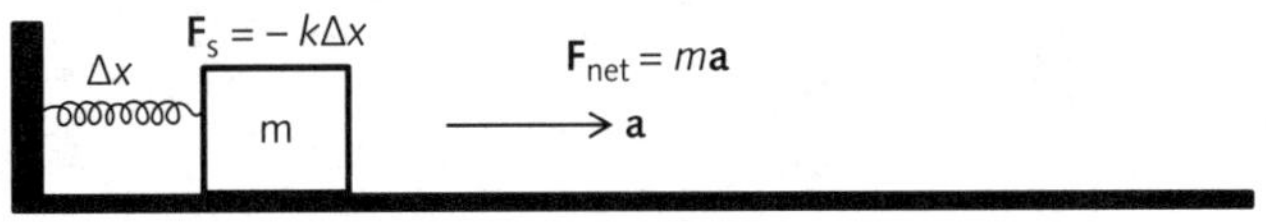

Figure 7.4 Mass oscillating on a spring

Problem Calculate the period of a 0.3-kg mass attached to a spring that has a spring constant $k = 25$ N/m.

Solution

Step 1. We have all of the given information, and so we can simply substitute in our given values (with units) to solve for the period:

$$T = 2\pi\sqrt{\frac{0.3 \text{ kg}}{25 \text{ N/m}}} = 0.69 \text{ s}$$

Consider now a simple pendulum that consists of a small mass attached to a string. When the mass is pulled to one side, the pendulum swings back and forth. This motion approximates simple harmonic motion (see Figure 7.5).

Because all masses fall with the same acceleration ($\mathbf{g}$), the period is independent of the mass. The period of the motion is given by:

$$T_{\text{pendulum}} = 2\pi\sqrt{\frac{L}{\mathbf{g}}} \text{ (period varies directly with square root of length)}$$

Problem A simple pendulum with a 0.3-kg mass is brought to an imaginary planet. The pendulum is 0.6 m long and is observed to have a period of 0.65 s. What is the value of $\mathbf{g}$ on this planet?

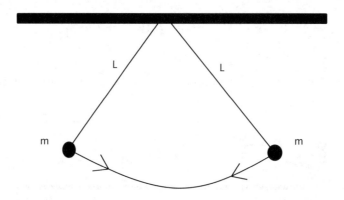

Figure 7.5 Simple pendulum

Solution

Step 1. The period of the pendulum is independent of the mass but does depend on the acceleration due to gravity. We can rearrange the formula to solve for **g**:

$$\mathbf{g} = 4\pi^2 L/T^2$$

Step 2. Now substitute in the given values (with units):

$$\mathbf{g} = 4\pi^2(0.6 \text{ m})/(0.65 \text{ s})^2 = 56.06 \text{ m/s}^2$$

Torque and Rotational Equilibrium

Consider a meterstick suspended from a string (see Figure 7.6). Two different masses are attached to it so that it balances at the 50 cm mark. This balance point (sometimes called a *fulcrum*) is also known as the *center of mass* (or *center of gravity* in this case). The system is in static equilibrium (since it is not moving), but something else is going on.

Because the masses are not equal, and the force of gravity is acting downward on both, we cannot state that equilibrium is caused only by the sum of all forces being equal to zero (that would be correct at the balance point, where the combined effect of the weights downward is balanced by the upward support force of the string). If, however, one of the masses were moved, the system would rotate (or *torque*) out of equilibrium. This torque is a rotation that results when a force acts at a right angle to a *lever arm* (like turning a wrench to loosen a bolt). When the meterstick is balanced, we say that it is in *rotational equilibrium* and the sum of all *torques* is equal to zero.

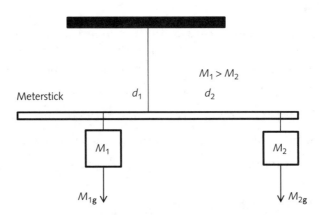

Figure 7.6 Rotational equilibrium

For each mass, the force of gravity ($\mathbf{F}_g = m\mathbf{g}$) acts through the lever arm distance (d), which is measured from the balance point. The *torque* is defined by the equation:

$$\text{Torque} = \text{force} \times \text{lever arm distance}$$

The mass M_1 tries to produce a counterclockwise torque, while the mass M_2 tries to produce a clockwise torque. If these two torques are equal and opposite, then the system will be in rotational equilibrium (even if $M_1 > M_2$). Newton's Laws of Motion apply to rotational dynamics as well as linear (or translational) dynamics. We could think of a *rotational inertia* (sometimes called the *moment of inertia*) as the tendency of an object to resist a torque. It would depend on the mass of the object and the distribution of the base relative to an axis of rotation (we will not discuss rotational dynamics in this book).

In the same way, the center of mass is very important when considering compound objects or systems of interacting objects. The *center of mass* is a mathematical point where all of the mass can be considered to be concentrated. A net force applied to the center of mass will *not* produce a torque.

Problem   A 0.5-kg mass is suspended from a meterstick. It is used to balance an unknown mass on the other side of a suspension string. If the 0.5-kg mass is located 0.35 m from the balance point, and the unknown mass is located 0.15 m from the balance point (on the other side), what is the value of the unknown mass?

Solution

Step 1. This problem illustrates the common example of a *balance scale*, which is used in many cultures throughout the world. It involves the principle of rotational equilibrium and balanced torques:

Sum of all clockwise torques = sum of all counterclockwise torques

Step 2. Identify all torques on either side of the balance point:

$M_1\mathbf{g}d_1 = M_2\mathbf{g}d_2$ (notice that the actual value of $\mathbf{g}$ is irrelevant!)

Step 3. Substitute in the given values (with units) and solve for the missing mass:

$(0.5 \text{ kg})\mathbf{g}(0.35 \text{ m}) = M_2\mathbf{g}(0.15 \text{ m})$

$M_2 = 11.17 \text{ kg}$

Exercise 7.1

1. A 0.2-kg mass is attached to a string and twirled overhead in a horizontal circle. The string has a radius of 0.5 m and the mass is observed to make 20 revolutions in 30 s. The plane of the circle is approximately 2.2 m above the ground.

   a. What is the period of the uniform circular motion?

   b. What is the magnitude of the tangential velocity of the mass?

   c. What is the magnitude of the centripetal force acting on the mass?

   d. If the mass is released, it will become a horizontally launched projectile. How far (horizontally) from the release point will it land?

2. A projectile is launched from the ground with a velocity of 125 m/s at a 25° angle to the horizontal. Calculate the maximum height and horizontal range of the projectile.

3. What must be the length of a simple pendulum (on earth) so that its period is equal to 1 s?

4. A suspended meterstick is in rotational equilibrium. If a 1-kg mass is placed 0.15 m from the balance point, where should a 0.4-kg mass be placed (relative to the balance point) to maintain rotational equilibrium?

8

Work, Energy, and Power

In this chapter you will learn about the concept of energy. Newton's Laws of Motion dealt with the concept of force vectors. Here we will explore the scalar concept of energy and learn that forces are just the agents by which energy is transferred. The mechanism for this process is called *work*. The Law of the Conservation of Energy is one of the most fundamental concepts in all of physics.

Work and Power

Consider an applied force that is acting on a mass (Figure 8.1). If this is a net force, then the mass will accelerate according to Newton's second law:

$$\mathbf{F} = m\mathbf{a}$$

Now recall from our study of motion that

$$\mathbf{v}_f^2 - \mathbf{v}_i^2 = 2\mathbf{a}\mathbf{d}$$

and

$$\mathbf{a} = \frac{\mathbf{v}_f^2 - \mathbf{v}_i^2}{2\mathbf{d}}$$

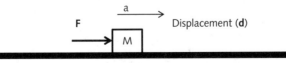

Figure 8.1 An applied force acting on a mass

If we substitute the expression for the acceleration back into Newton's second law, we obtain (after a little rearranging):

$$Fd = \frac{1}{2}mv_f^2 - \frac{1}{2}mv_i^2$$

The quantity on the left (Fd) is called the *work done to the object*, and the quantity on the right is called the *change in the kinetic energy* ($\text{KE} = \frac{1}{2}mv^2$). These are scalar quantities since it does not matter whether the velocity is positive or negative. The concept of work is very important in physics and it is related to the transfer of energy. We will discuss kinetic energy later on in the chapter.

The units of work are N · m, which are also known as *joules* (J). This unit is named for James Prescott Joule, a nineteenth-century physicist who investigated the relationship between heat and mechanical work.

Using the formula $W = Fd$, we can see that for a constant force acting on a mass (neglecting friction), a graph of force vs. displacement would be like Figure 8.2.

The work done on the mass is equal to the *area under the graph* of force versus displacement. In this case, the work done by the 10-N force acting through a displacement of 5 m is equal to 50 J. If an object is being *held*, then *no* work is being done. If kinetic friction ($\mathbf{f}_k$) is present, then we can consider the net work is being done by an applied force and friction ($W_f = -fd$). If an object has a net force of zero (moving at a constant velocity) but has an applied force acting on it, then the *net* (or total) work is equal to zero, since the work done by friction cancels out the work done by the applied force.

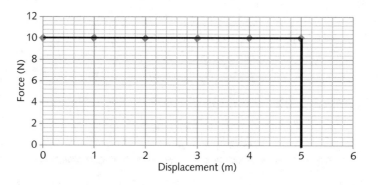

Figure 8.2 Force vs. displacement

Misconception

It is important to remember that work and energy are *scalar* quantities (both measured in units of joules). Mathematically, it may appear that we are multiplying two vectors (force × displacement), but the mathematical operation being employed is beyond the scope of this book. You can consult a mathematics book on vector analysis to explore and fully understand the concept of what is called the *scalar dot product* (we will not discuss this here). This operation, however, does lead to a scalar quantity.

Also, the inclusion of the negative sign for the work done by friction is simply to remind us that friction will take energy (KE) out of the system (it is typically converted into heat). There *must* be a change in the energy of the system in order for work to be done. An object that is not accelerating is *not* having work done to it ($\Delta KE = 0$).

If the force is acting at an angle (see Figure 8.3), then to find the work done, we need to find the component of the applied force in the same direction as the acceleration.

The rate at which work is done is called *power*.

$$P = \text{power} = \text{work/time (units are joules/seconds or watts, W)}$$

Problem   A 20-kg mass is being pulled horizontally by a rope that makes a 30° angle to the horizontal. If the force in the rope is 100 N (neglect friction), calculate the work done if the mass is pulled 10 m and the power generated if the work is done over a period of 2 minutes.

Solution

Step 1. To calculate the work done, we need the appropriate component of applied force:

$$\mathbf{F}_x = \mathbf{F} \cos \theta = (100 \text{ N})\cos(30°) = 86.6 \text{ N}$$

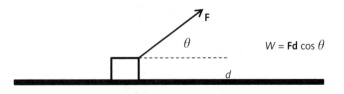

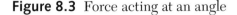

Figure 8.3 Force acting at an angle

Easy Physics Step-by-Step

Step 2. Since the applied force is the net force (in the horizontal direction), then

$$W = \mathbf{F}d = (86.7 \text{ N})(10 \text{ m}) = 866 \text{ J}$$

Step 3. To find the power, we need to recognize that the time period must be in seconds:

$$t = 2 \text{ min} = 120 \text{ s}$$
$$P = (866 \text{ J})/(120 \text{ s}) = 7.22 \text{ W}$$

Work Done by Springs

If an applied force is variable, then the easiest way to calculate the work done is to find the area under a graph of force versus displacement. If the graph is curved, then there will be difficult approximations for the area (which are beyond the scope of this book). However, if the force graph involves idealized lines, then the area (just like the velocity versus time graphs) can be divided up into sections of rectangles and triangles.

In the case of a spring, the force is variable, but linear according to Hooke's Law:

$$\mathbf{F} = kx \text{ (where } x \text{ is the amount of elongation)}$$

A graph of Hooke's Law would be a line that starts from the origin (see Figure 8.4).

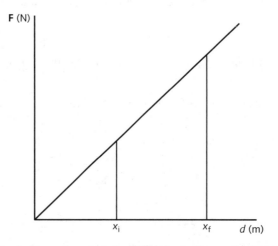

Figure 8.4 A graph of Hooke's Law for finding work done on a stretched spring

The area under the graph (the work), from some initial length (x_i) to some final length (x_f), is equal to the difference between the area of a large triangle (from $x = 0$ to $x = x_f$) and the area of a small triangle (from $x = 0$ to $x = x_i$). The area of a triangle is $\frac{1}{2}$ (base × height), and recalling that $F = kx$, we obtain:

$$W = \frac{1}{2}kx_f^2 = \frac{1}{2}kx_i^2$$

Problem Shown is a graph of force vs. displacement for a stretched spring. How much work was done to stretch the spring from $x = 0.2$ m to $x = 0.4$ m?

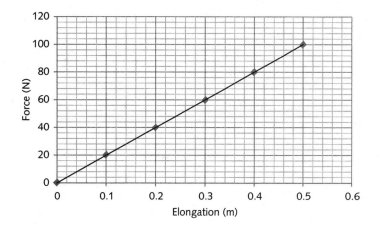

Solution

Step 1. The goal is to find the work done needed to stretch the spring from $x = 0.2$ m to $x = 0.4$ m, so we need to find the area of that segment under the graph shown earlier. Another method would involve finding the value of k and then using the equation derived for the work done to a spring.

Step 2. The slope of the line is $k = 200$ N/m. But the area of the region in question is

$$W = \frac{1}{2}(80\text{ N})(0.4\text{ m}) - \frac{1}{2}(40\text{ N})(0.2\text{ m}) = 12\text{ J}$$

Step 3. If we use the equation derived earlier,

$$W = \frac{1}{2}(200 \text{ N/m})(0.4 \text{ m})^2 - \frac{1}{2}(200 \text{ N/m})(0.2 \text{ m})^2 = 12 \text{ J}$$

Simple Machines

Consider a mass sliding down a frictionless incline from rest (Figure 8.5). The length of the incline is d and the height is h. We already know that the force parallel to the incline is given by: $F\| = m\mathbf{g}\sin\theta$

The work done by the component of gravity parallel to the incline is given by

$$W\| = m\mathbf{g}d \sin \theta$$

If the mass were dropped from the height h, the work done by gravity (directly) would be

$$W\mathbf{g} = m\mathbf{g}h$$

In the absence of friction, the work done by both of these paths would be equal:

$$W\| = W_g$$

and

$$h = d \sin \theta$$

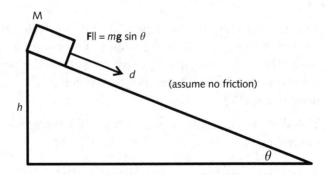

Figure 8.5 An incline being used as a simple machine

In this ideal situation, the force given by gravity ($m\mathbf{g}$) is greater than the force parallel to the incline ($m\mathbf{g}\sin\theta$). However, the length of the incline is proportionally larger than the height.

Thus, while the work done is the same, the force can be reduced at the expense of distance. This is the essence of a *simple machine*. If we reverse the process (push the object up the ramp), we can see the advantage to this machine. We can apply less effort (force) to do the same work as lifting the object to the height h if we compensate by going a longer distance. The height h is considered the *resistance distance* that must be moved through against the force of gravity.

The Ideal Mechanical Advantage (IMA) is defined to be equal to

$$\text{IMA} = \frac{\text{effort distance}}{\text{resistance distance}} = \frac{\text{length of incline}}{\text{height of incline}}$$

For example, if the IMA of a ramp is equal to 3 (length is three times as long as the height), then under ideal conditions (no friction) we can use one-third the weight of the object as effort when pushing it up the ramp (as long as we go three times as far).

If there is friction, the IMA does not change, but we would need to use more effort to overcome friction. The *real mechanical advantage* would be defined as the ratio of resistance force (in this case, the weight) to the actual effort force used:

$$\text{RMA} = \frac{\text{resistance force}}{\text{effort force}}$$

The work input (W_i) is defined to be:

$$W_i = \text{effort force} \times \text{effort distance}$$

And the work output (W_o) is defined to be:

$$W_o = \text{resistance force} \times \text{resistance distance}$$

The efficiency of the machine is defined to be:

$$\%\text{ efficiency} = \frac{W_o}{W_i} \times 100 = \frac{\text{RMA}}{\text{IMA}} \times 100$$

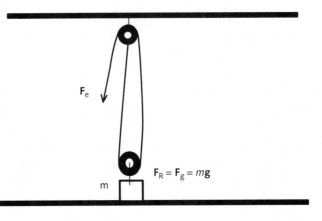

Figure 8.6 Pulley system

If the work input is equal to the work output, then the machine is 100% efficient and no energy is lost to friction. Any difference between work input and work output is lost as the work done by friction:

$$W_f = W_o - W_i$$

Another common simple machine is a pulley system (see Figure 8.6). In this pulley system, the resistance force ($\mathbf{F}_R$) is given by the weight ($m\mathbf{g}$) and the effort force ($\mathbf{F}_e$) is provided by the person pulling on the string. The IMA for this system is equal to 2, because the two supporting strings each support half the load. The effort distance would be equal to the amount of string pulled down by the person, while the resistance distance would be equal to the height to which the mass is raised. If the IMA = 2, then effort distance will be twice the resistance distance and the effort force will be equal to half the weight of the mass.

Problem In a pulley system, a 50-kg mass is lifted a distance of 5 m by pulling 20 m of rope. It is observed that 150 N of effort force is needed to accomplish this task. Calculate the IMA, RMA, and the efficiency for this pulley system.

Solution

Step 1. We observe that the effort distance is equal to the amount of rope pulled by the person and the resistance distance is equal to the height to which the mass is raised. Thus,

IMA = 20 m/5 m = 4

Step 2. The IMA is equal to 4, which means that, ideally, we should need to use as effort only one-fourth of the weight. A 50-kg mass weighs 490 N (in magnitude; this is the resistance force); therefore, the ideal effort should be equal only to 122.5 N. The actual effort used is 150 N. Thus,

RMA = 490 N/150 N = 3.27

Step 3. The efficiency of this pulley system is obtained by:

% efficiency = (3.27/4.00) × 100 = 82%

This means that the approximately 18% of the remaining energy is used to overcome friction.

Potential Energy and Kinetic Energy

According to our definition, no work is being done while holding a mass up above the ground (no change in *kinetic energy*). However, work was done against gravity while the mass was being lifted to that height (see Figure 8.7). The work done against gravity would be given by:

$$\mathbf{Wg} = m\mathbf{g}\Delta h = m\mathbf{g}\,(h_2 - h_1)$$

If the mass were released from that height, gravity would now do work in bringing the mass down to the ground. While it is being held, the potential exists for work to be done. Since energy is defined as the *capacity to do work*, we define the *potential energy* of a mass to be equal to the energy possessed by the mass due to separation from the earth (relative to an arbitrary base level):

$$PE = m\mathbf{g}\Delta h \text{ (units measured in joules)}$$

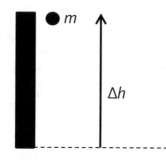

Figure 8.7 Work done against gravity

If we define the base level (h_1) as zero, then the potential energy is just given by $PE_g = mgh$. Thus, the work done (when lifting a mass) is just equal to the change in potential energy:

$$Wg = \Delta PE_g$$

The potential energy will increase when the mass is raised higher above the base level. For example, we can choose a table top to be the base level (PE = 0 relative to the table) and measure the PE relative to that position.

A spring that has been stretched (or compressed) also has potential energy stored within it (*elastic potential energy*):

$$\Delta PE_{spring} = \frac{1}{2}kx_f^2 - \frac{1}{2}kx_i^2$$

We have already defined the *kinetic energy* (energy of a mass in motion) as

$$KE = \frac{1}{2}mv^2 \left(\text{units are in joules}\right)$$

Notice that as a mass falls, it accelerates. Thus, we can observe that the potential energy converts into kinetic energy. In the absence of any air resistance, we would have:

$$\Delta PE = \Delta KE$$

This is called the *conservation of mechanical energy*. Energy (like work) is a scalar quantity. The conservation of energy means that energy is transferred from one form to another through the process of work being done.

Conservation of Energy

Energy is never created nor destroyed. This fundamental principle of nature is referred to as the Law of the Conservation of Energy. In this chapter we have restricted ourselves to mechanical (potential and kinetic) energy only. However, there are many forms of energy, such as mechanical, thermal, electromagnetic, chemical, and nuclear. If we restrict ourselves to only mechanical energy, then we can state that while a system is changing:

Total mechanical energy before = total mechanical energy after
$$PE_i + KE_i = PE_f + KE_f$$

Problem A 0.3-kg mass is dropped from a height of 100 m. How much potential energy did it initially have? How much kinetic energy will it have as it hits the ground? What will be the velocity of the object as it hits? (Assume no air resistance.)

Solution

Step 1. The starting potential energy (relative to the ground in which PE = 0) is equal to:

$$PE_g = mgh = (0.3 \text{ kg})(9.8 \text{ m/s}^2)(100 \text{ m}) = 294 \text{ J}$$

Step 2. Because there is no air resistance, we have:

$$\Delta PE \text{ (loss)} = \Delta KE \text{ (gain)} = 294 \text{ J of KE gained as it hits the ground}$$

Step 3. As the mass hits the ground, all of the potential energy is converted into kinetic energy:

$$KE_{bottom} = 294 \text{ J} = \frac{1}{2}(0.3 \text{ kg})\mathbf{v}^2$$

$$\mathbf{v} = -44.3 \text{ m/s (downward direction)}$$

Problem A 50-kg cart is resting on top of a frictionless track that is 20 m high (as shown). At the end of the track, which is on the ground, is a spring with $k = 500$ N/m. When the mass is released the cart will slide down the track and impact the spring. How much will the spring compress when the mass finally stops?

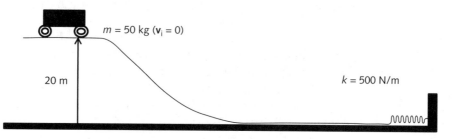

Solution

Step 1. The goal of this problem is to find the compression of the spring. However, to find that value, we first need to recognize that what we are really dealing with is the conservation of mechanical energy. The cart begins only with gravitational potential energy (it is at rest) and converts that into kinetic energy as it slides down the

frictionless hill. Along the bottom, all of the energy is kinetic energy (no PE relative to the ground) until it impacts the spring. The work needed to compress the spring comes from the loss of KE (which converts into elastic potential energy).

Step 2. Calculate the PE at the top of the hill, which will be lost to KE at the bottom:

$$\Delta PE \text{ (loss)} = (50 \text{ kg})(9.8 \text{ m/s}^2)(20 \text{ m}) = 9800 \text{ J} = \Delta KE \text{ (gain)}$$

Step 3. This gain in KE converts into elastic potential energy for the spring ($PE_i = 0$):

$$9800 \text{ J} = \frac{1}{2}(500 \text{ N/m})x^2$$

$x = 6.3 \text{ m}$

The conservation of energy is a very important part of learning physics. It starts from the concept of mechanical work and ends with the triumph of Albert Einstein's mass-energy equivalence equation $E = mc^2$ (where c = velocity of light: 3×10^8 m/s). We will encounter this mass-energy equivalence at the end of our journey when we study nuclear energy. In the meantime, energy is an abstract but vital component to the understanding of the mechanical universe.

Problem-Solving Strategies to Avoid Missteps

As we have mentioned previously, it is important to remember that work and energy are *scalar* quantities. Work is the means by which energy is transferred and energy is the capacity to do work. When setting up a problem, make sure you refer to the problem-solving ring. It may be that problems are not set up in a sequential a,b,c format. It is important to identify the forces and types of energies involved. For simple machines, under ideal conditions, work input is equal to work output. If friction is present, then the work done by friction will reduce kinetic energy. The inefficiency of a machine is a reflection of this energy loss. Make sure to write out all equations, identify all given information, and write down your general statements about the conservation of energy. For problems involving work, make sure you have identified the correct force component for the calculations.

Exercise 8.1

1. A 15-kg mass is pushed up a ramp that is 2.5 m long and 1.25 m high.

   a. How much work would be done if the mass were to be lifted to the height of 1.25 m?

   b. What is the IMA of the ramp?

   c. It is observed that a person needs 100 N of force to push the mass up the ramp. How much work is done by the person?

   d. How much work is used to overcome friction?

   e. What is the efficiency of the ramp?

2. A motor generates 40 kilowatts (40 kW) of power to lift a 1000-kg mass.

   a. How much work is done in 2 minutes?

   b. What was the average force applied if the distance covered during that time was 100 m?

3. A 100-kg cart is moving at 5 m/s on top of a 12-m hill, as shown. What is the velocity of the cart when it is on top of the 7-m hill? Neglect any friction.

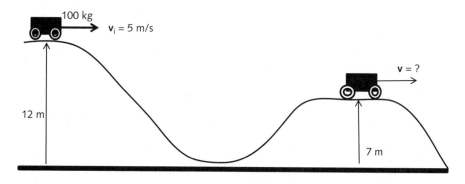

4. A 0.5-kg mass is sitting on top of a compressed spring ($k = 20$ N/m). The spring has been compressed by 0.5 m. When released, the mass will be projected vertically up into the air. How high will the mass go?

9

Momentum

In this chapter you will learn about the *Law of the Conservation of Momentum*. This law, along with the Law of the Conservation of Energy, is a vital part of our journey in exploring the mechanical universe. When matter interacts, usually through collisions, momentum is always conserved.

Impulse and Momentum

You will recall from Chapter 6 that Newton's second law ($\mathbf{F}_{net} = m\mathbf{a}$) can be written as:

$$\mathbf{F}t = m\Delta\mathbf{v} = m\mathbf{v}_f - m\mathbf{v}_i$$

for a constant applied force. The quantity on the left-hand side is defined as the impulse, and the quantity on the right is defined as the change in momentum.

$$\text{Impulse} = \mathbf{J} = \mathbf{F}t = \text{force} \times \text{time (units are N} \cdot \text{s)}$$
$$\text{Momentum} = \mathbf{p} = m\mathbf{v} = \text{mass} \times \text{velocity (units are kg} \cdot \text{m/s)}$$

Both of these quantities are vectors and so once again direction plays a very important role. In fact, using this definition we can think of Newton's second law as:

$$\mathbf{F}_{net} = (\Delta\mathbf{p})/(\Delta t)$$

In other words, *the net force acting on a mass is equal to the rate of change of its momentum.*

Problem   A 2-kg mass has a constant force of 10 N acting on it for 10 s. If the initial velocity was 5 m/s, what is the final velocity of the mass?

Solution

Step 1. In this case, we are using the concept of impulse and change in momentum.

$$\mathbf{F}t = m(\mathbf{v}_f - \mathbf{v}_i)$$

Step 2. Substitute in the given values with units:

$$(10 \text{ N})(10 \text{ s}) = (2 \text{ kg})(\mathbf{v}_f - 5 \text{ m/s})$$

$$\mathbf{v}_f = 55 \text{ m/s}$$

The impulse can also be determined by looking at a graph of net force versus time.

Problem   A 2-kg mass with an initial velocity of 5 m/s has a constant net force acting on it as shown in the graph. What is the impulse acting on the mass during the 5-s interval? What is the final velocity of the mass after the 5-s interval?

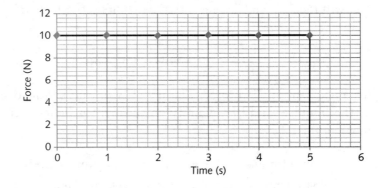

Solution

Step 1. The impulse after 5 s would be equal to the area of the rectangle:

Total impulse = total area = $(10 \text{ N})(5 \text{ s}) = 50 \text{ N} \cdot \text{s}$

Step 2. Now we know that:

Impulse = change in momentum = $m\Delta\mathbf{v} = m(\mathbf{v}_f - \mathbf{v}_i)$

$$50 \text{ N} \cdot \text{s} = (2 \text{ kg})(\mathbf{v}_f - 5 \text{ m/s})$$

$$\mathbf{v}_f = 30 \text{ m/s}$$

Problem A graph of net force versus time is shown for a 5-kg mass moving horizontally:

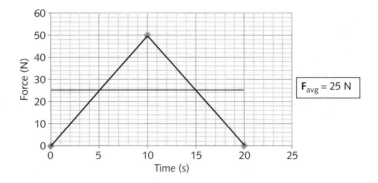

If the mass initially starts from rest, what is its final velocity after 20 s?

Solution

Step 1. The goal is to find the final velocity of the mass after 20 s. However, to determine that velocity, we first need to find the impulse acting on the mass. This impulse is equal to the area under the graph (which is in the shape of a triangle during the 20-s interval).

Step 2. The impulse = area of a triangle = $\frac{1}{2}bh = \frac{1}{2}(20\text{ s})(50\text{ N}) = 500\text{ N} \cdot \text{s}$

Step 3. The final velocity can now be calculated:

Impulse = change in momentum

$500\text{ N} \cdot \text{s} = (5\text{ kg})(\mathbf{v}_f - 0\text{ m/s})$

$\mathbf{v}_f = 100\text{ m/s}$

Notice that the impulse of 500 N · s could have been found if we had a constant net force of 25 N acting on the mass for the entire 20-s interval. This force would be the value of the *average net force*:

Average net force = total impulse/total time = total area/total time

Misconception

When dealing with momentum, we often define an *isolated system*. This can be thought of as two or more interacting masses (with no *external* forces acting). The impulse is *not* equal to the *momentum*, but it is equal to the *change in momentum*. The units N · s are equivalent to the units kg · m/s.

Conservation of Momentum and One-Dimensional Collisions

When one object impacts another, momentum is transferred. When momentum is transferred, forces are involved. If we think of Newton's third law (*for every action there is an equal but opposite reaction*), then the situation seen in Figure 9.1 takes on the following implications. Suppose mass M_1 has a velocity $\mathbf{v}_1$ and it impacts with a mass M_2 moving with a slower velocity $\mathbf{v}_2$.

When the two masses collide, the forces of action and reaction act simultaneously. This means that because the forces are equal and opposite, then the impulses applied to each mass are equal and opposite. If the impulses are equal and opposite, that means the *changes* in momentum (for each mass) are equal and opposite. If the *changes* in momentum are equal and opposite, this means that the total change in momentum for the system of the two masses is equal to zero. In other words, *for an isolated system (in the absence of external forces), the total linear momentum of a system remains the same*. This statement is known as the *Law of Conservation of Linear Momentum*. This can also be written as:

Total momentum before collision = total momentum after collision
$$M_1\mathbf{v}_{1i} + M_2\mathbf{v}_{2i} = M_1\mathbf{v}_{1f} + M_2\mathbf{v}_{2f}$$

Although we will not discuss it in detail here, an object that is rotating has *angular momentum*. We can state the *Law of the Conservation of Angular Momentum* as follows: *in the absence of any external torques, the angular momentum of a system is conserved*. This can be easily observed when a spinning ice skater brings in her arms and then spins faster.

Elastic Versus Inelastic Collisions (One Dimension)

Momentum is a vector quantity. The conservation of momentum is satisfied in more than one dimension, but we will be considering only one-dimensional collisions.

Figure 9.1 Transfer of momentum

When masses collide, kinetic energy (horizontally, in one dimension—we will ignore potential energy changes) is transferred. If the kinetic energy remains the same, then the collision is called *elastic*. If the kinetic energy is changed (typically, some KE is lost), then the collision is called *inelastic*. If the masses stick together, then we say that we have a *completely inelastic collision*. In all of these situations, the linear momentum is conserved.

Problem A 0.2-kg steel ball is moving horizontally at 5 m/s. It collides with a 0.4-kg steel ball at rest. If the velocity of the 0.4-kg ball is 2 m/s after the collision, what is the new velocity of the 0.2-kg ball? Is this collision elastic?

Solution

Step 1. The goal is to find the final velocity of the first mass (the second mass is initially at rest). We write:

Total momentum before = total momentum after

Step 2. Substitute in the given values and solve for the final velocity of the 0.2-kg mass:

$(0.2 \text{ kg})(5 \text{ m/s}) + (0.4 \text{ kg})(0 \text{ m/s}) = (0.2 \text{ kg})\mathbf{v}_{1f} + (0.4 \text{ kg})(2 \text{ m/s})$

$\mathbf{v}_{1f} = 1 \text{ m/s}$

Step 3. If the collision is elastic, then the kinetic energy would be conserved. Thus,

$$\text{Total KE}_i = \frac{1}{2}(0.2 \text{ kg})(5 \text{ m/s})^2 + 0 = 2.5 \text{ J}$$

$$\text{Total KE}_f = \frac{1}{2}(0.2 \text{ kg})(1 \text{ m/s})^2 + \frac{1}{2}(0.4 \text{ kg})(2 \text{ m/s})^2 = 0.9 \text{ J}$$

The kinetic energy before is not equal to the kinetic energy after, so the collision is *not* elastic (even though the masses remain separate).

Problem A 100-kg cart moving at 20 m/s collides and sticks together with an identical 100-kg cart that is at rest. What is the final velocity of the system after the collision?

Solution

Step 1. This is a *completely inelastic collision* because the masses remain joined together. The momentum is still conserved:

Total momentum before = total momentum after

Step 2. Since the masses join together, they move as one combined mass after colliding:

$$(100 \text{ kg})(20 \text{ m/s}) + (100 \text{ kg})(0 \text{ m/s}) = (200 \text{ kg})\mathbf{v}_f$$

$$\mathbf{v}_f = 10 \text{ m/s}$$

The applications of conservation of momentum are widespread in physics. Momentum is conserved in all dimensions of space (and even at very high velocities, where Einstein's equations of *special relativity* take over). Momentum is conserved on the atomic level. The collisions of molecules in an ideal gas lead to the *kinetic theory of gases* (and concepts such as Boyle's and Charles' laws).

Make sure you write down all of the given information, write out the equations for the conservation of momentum (before and after the collision), and always include the units. Remember, impulse is equal to the change in momentum (not the momentum itself) and the units of $N \cdot s$ are equivalent to the units $kg \cdot (m/s)$.

Problem-Solving Strategies to Avoid Missteps

Solving problems with momentum can sometimes be difficult, so you should always consult the problem-solving ring for strategies. It is important for you to understand that in an isolated system (of two or more objects), momentum is always conserved. Kinetic energy is conserved only in *elastic* collisions. *Inelastic* collisions do not necessarily mean that the objects stick together.

Exercise 9.1

1. A 3-kg mass, with an initial velocity of 10 m/s, is acted upon by a variable net force. The graph of force versus time is shown.

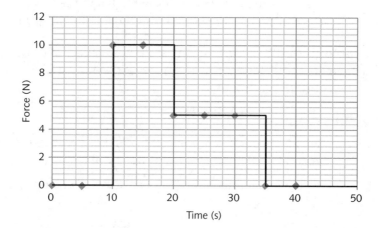

a. What is the initial momentum of the mass?

b. What is the total impulse applied to the mass during the 50-s time interval?

c. What is the final velocity of the mass after 50 s?

2. A 4-kg mass moving to the right with a velocity of 2 m/s collides with a 2-kg mass that is at rest.

a. If the 4-kg mass has a velocity of 0.66 m/s (to the right) after the collision, find the new velocity of the 4-kg mass.

b. Was the collision elastic or inelastic?

3. A 10-kg mass moving to the right with a velocity of 8 m/s collides, and then sticks together, with a 5-kg mass moving to the left at 6 m/s.

a. What is the new velocity of the combined mass?

b. How much kinetic energy was lost in the collision?

10

Gravitation

In this chapter you will learn about Newton's *Law of Gravitation* and how the universal law of gravitational force applies to all bodies in the universe. This is a significant point in the steps we have taken on our journey. Newton's understanding (along with the discoveries of Johannes Kepler and Galileo) shows that the same forces that work on the earth operate everywhere, from the moon to the farthest galaxy.

Kepler's Laws of Planetary Motion

From the earth's frame of reference, the sun, moon, and stars appear to make a circular orbit around it (rising in the east and setting in the west). To the ancient Greeks (especially to the philosopher Plato), circular motion was natural and represented perfection. From Newton's Laws we now know that the tendency of all objects is to maintain constant velocity (motion in a straight line at constant speed). Circular motion is the result of a net force acting at right angles to the instantaneous velocity.

From the earth (which to the ancients appeared to be fixed), five objects could be observed with the naked eye to wander against the background of the "fixed" stars. The ancient Greeks called them *planets* (meaning "wanderers"). Two of them (Mercury and Venus) always appeared to be near the sun (either rising an hour or two before sunrise or setting an hour or two after sunset). The others (Mars, Jupiter, and Saturn) sometimes appeared to make loops following the plane of the ecliptic. The *ecliptic* is the apparent path the sun takes in the sky against the background of twelve groupings of stars known as the *zodiacal constellations*.

Ptolemy of Alexandria (in the first century CE) developed an elaborate mathematical system of *epicycles* (essentially, circles rotating around other circles), that would maintain the central position of the earth and account for the seemingly erratic motion of the then five known planets. Nicholas Copernicus (1473–1543) in his book *The Revolution of Heavenly Orbs*, published in 1543, mathematically demonstrated that the system of planetary motion could be better explained if the sun were placed at the center and the other planets orbited the sun at different distances and velocities.

Johannes Kepler (1571–1630), using observations of Tycho Brahe (1546–1601), synthesized the motion of the planets into three laws (see Figure 10.1).

Kepler's Three Laws of Planetary Motion

1. The planets orbit the sun in elliptical paths with the sun at one focus.

2. A planet sweeps over equal areas in equal time. The planet moves faster when it is closer to the sun and slower when it is farther away.

3. The ratio of the cube of the mean length of the semimajor axis of the ellipse to the square of the orbital period is equal to a constant for all of the planets ($R^3/T^2 = k$).

You should recall from geometry that an ellipse differs from a circle. A circle is the shape formed from the set of all points equidistant from a fixed point. With an ellipse, there are two fixed points (called the *foci*; a single point is a *focus*), and from any point along the ellipse, the sum of the lengths of the lines from each of the foci to that point is a constant. The two foci lie along the major axis of the ellipse. The portion of that axis measured from

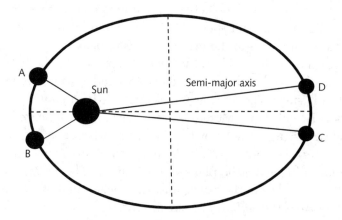

Figure 10.1 Illustration of Kepler's three laws

the center is called the semimajor axis. If the length of the semimajor axis is a, then the distance from the center of the ellipse to a focus is some fraction $e \cdot a$, where the number e represents the *eccentricity* of the ellipse (that is, the degree to which it is noncircular).

The sketch in Figure 10.1 is not drawn to actual scale. The eccentricity is exaggerated for the purpose of illustration. The actual eccentricities of the planetary orbits are quite small (less than 10%), and the mean length of the semimajor axis is approximately the same as the radius of a nearly circular orbit (the orbital eccentricity of the moon's orbit is very small). In the diagram, if the time for the planet to go from point A to point B is the same as the time for it to go from point C to point D, then the areas of the sectors (pie shapes) swept out by the planet are equal (according to Kepler's second law).

As we shall see, Newton was able to derive Kepler's laws using his concepts of dynamics, and that enabled him to determine many of the parameters needed to eventually launch human beings to the moon three centuries later.

Newton's Law of Gravitation

Kepler's three laws apply to all orbiting systems, but the constant k is not universal. This means that although $R^3/T^2 = k$ for the planets orbiting the sun and the moons of Jupiter also obey the rule $R^3/T^2 = k$, the value of k for that system is different from the constant for the planets orbiting the sun.

In his *Principia* (published in 1687), Newton derived a more universal law based on the force interaction between two isolated masses. In modern notation (and units). Newton's law states:

$$\mathbf{F}_g = G \frac{m_1 m_2}{r^2}$$

where $G = 6.67 \times 10^{-11} \ \mathrm{N \cdot m^2/kg^2}$; that is, *there is an attractive force between any two masses that is directly proportional to the product of the masses and inversely proportional to the square of the distances between the centers of the objects.*

In 1795, Henry Cavendish (1731–1810), using a *torsion balance* (two masses suspended from a thin wire), determined the value of Newton's universal gravitational constant G. It is important to note that for easier approximations we will only be considering *point* masses in this book. The situation becomes more complicated for irregularly shaped objects or if there are more than two objects present.

Problem   Calculate the gravitational force between a 2000-kg mass and a 5000-kg mass separated by a distance of 2.5 m. What would happen to the force if the distance between the masses was doubled?

Solution

Step 1. The goal of this problem is to find the force of gravitational attraction between two masses. All the information is given, so we just need to substitute in our given values with units.

Step 2. $\mathbf{F}_g$ = (6.67 × 10⁻¹¹ N·m²/kg²)(2000 kg)(5000 kg)/(2.5 m)² = 1.07 × 10⁻⁴ N

Step 3. Because the force is an inverse square law relationship, doubling the distance will have the effect of reducing the magnitude of the force by one-fourth (it will be quartered).

Misconceptions

It is important to remember that the distance between the masses needs to be squared in the calculation process. Also, as an *inverse square law*, the force gets weaker inversely to the square of the distance between the masses. As we have just seen, if the distance is doubled, the force is reduced by four times. If the distance were tripled, then the force would be reduced by nine times. However, if the distance were halved, then the force would *increase* by four times. We will encounter an inverse square law again when we discuss electrostatic forces in Chapter 16.

Gravitation is a force that always attracts two masses. Newton himself never discussed the origin of gravity. His law only describes how it works (and in an imaginary system of only two objects). In Newton's own words (in Latin), "*Hypotheses non fingo*" ("I frame no hypotheses").

You will recall from our discussion of kinematics (Chapter 4) that Galileo demonstrated that in the absence of any external forces, all objects fall (near the surface of the earth) with the same acceleration ($\mathbf{g}$ = 9.8 m/s²). Newton was able to explain why this is correct using his Law of Gravitation. He was, it is said, inspired by watching apples fall from trees in his family's orchard.

Imagine an apple falling from a tree. The distance from the apple to the center of the earth is approximately the same as the radius of the earth (R_e = 6.37 × 10⁶ m). The force of gravity, using Newton's second law, would accelerate the falling apple (m_a) due to a net force:

$$\mathbf{F}_g = m_a \mathbf{g}$$

From the Law of Gravitation, the force can also be written as:

$$\mathbf{F}_g = G\frac{m_a M_e}{R_e^2}$$

where M_e = mass of the earth. Because these two forces must be identical, the mass of the apple is irrelevant and we see that:

$$\mathbf{g} = G\frac{M_e}{R_e^2}$$

If we use the experimentally verified value of $\mathbf{g}$ = 9.8 m/s² and the known radius of the earth, then we see that the mass of the earth is equal to an amazingly large M_e = 5.98 × 10²⁴ kg!

 Misconceptions

The value of $\mathbf{g}$ is the same for all falling objects (regardless of their mass) near the surface of the earth and in the absence of external forces. Air resistance, for example, would reduce the acceleration of falling until a *terminal velocity* (falling at constant velocity) is reached. This is what happens to a feather when it is dropped along with a penny. However, in many experiments, physicists can demonstrate that in a vacuum, a penny and a feather fall with the same acceleration!

The value (and meaning) of $\mathbf{g}$ is not really a constant (except very close to the surface of the earth) and is also conceptually more sophisticated. If we think of an object resting on a table, then its standard weight is given by $\mathbf{F}_g$ = $m\mathbf{g}$ (as expected). However, if we think of the force of gravity as a measured quantity, then the expression

$$\mathbf{g} = \mathbf{F}_g/m = 9.8 \text{ N/kg}$$

introduces us to the concept of a *gravitational field*.

This means that $\mathbf{g}$ is a measure of the strength of the gravitational field (in units of N/kg). A field (in the way it is used in physics) is a region of space where a force can be detected. In our case, if a test mass is brought near the earth, it will experience a force directed radially inward toward the center of the earth. Our previous expression for gravitation potential energy (PE$_g$ = mgh) in Chapter 8 implied a constant value for $\mathbf{g}$. We will encounter the concept of a field again when we study electric fields in Chapter 16.

Problem   Calculate the acceleration due to gravity near the surface of the moon given that

$$M_{moon} = 7.35 \times 10^{22} \text{ kg}$$

and

$$R_{moon} = 1.738 \times 10^{6} \text{ m}$$

What would be the weight of a 10-kg mass on the moon?

Solution

Step 1. The goal of this problem is find the value of **g** on the moon.

$$\mathbf{g}_{moon} = (6.67 \times 10^{-11} \text{ N} \cdot \text{m}^2/\text{kg}^2)(7.35 \times 10^{22} \text{ kg})/(1.738 \times 10^{6} \text{ m})^2$$

$\mathbf{g}_{moon} = 1.62 \text{ m/s}^2$ (this value is about one-sixth the value of **g** on the earth)

Step 2. A 10-kg mass would weigh only 16.2 N on the moon since

$$\mathbf{F}_{g} = m\mathbf{g} = (10 \text{ kg})(1.62 \text{ m/s}^2) = 16.2 \text{ N (it would weigh 98 N on earth!)}$$

Applications of Gravitation

Isaac Newton was able to explain all of Kepler's laws using his law of universal gravitation. We can explore some of these applications by looking at orbital motion. Let us assume an object is orbiting the earth in a nearly circular path (see Figure 10.2). Then, according to Newton's second law, the object is being acted upon by a net force (the centripetal force). In this case, the net force is due to gravity (as described by Newton's Law of Gravitation):

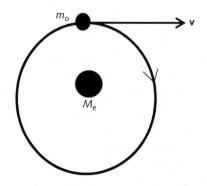

Figure 10.2 Object orbiting earth

$$\mathbf{F}_{net} = \mathbf{F}_{c} = m_{o}\mathbf{v}^2/R = m_{o}4\pi^2R/T^2 = \mathbf{F}_{g} = G\frac{m_{o}M_{e}}{R^2}$$

The velocity is tangent to the circle and is now referred to as the orbital velocity. We can calculate the orbital velocity from

$$m_{o}\mathbf{v}^2/R = G\frac{m_{o}M_{e}}{R^2}$$

$$\mathbf{v}_{orbit} = \sqrt{\frac{GM_{e}}{R}}$$

Kepler's third law can be derived using the other formula for centripetal force:

$$m_{o}4\pi^2R/T^2 = G\frac{m_{o}M_{e}}{R^2}$$

$$R^3/T^2 = GM_{e}/4\pi^2$$

Notice that both of these equations are independent of the mass of the orbiting object and depend only on the mass of the orbiting objects and the radial distance between them. They would be valid for any object orbiting around another object in a nearly circular path.

Problem Mimas is a satellite of the planet Saturn. It has a known orbital period of 8.14×10^4 s and a mean distance from Saturn equal to 1.86×10^8 m. Using this information, calculate the mass of the planet Saturn. Use your value for the mass of Saturn to calculate the orbital velocity of Mimas.

Solution

Step 1. The goal of this problem is to find the mass of Saturn using the orbital data observed about one of its satellites. We can rearrange Newton's version of Kepler's third law:

$M_{sat} = 4\pi^2R^3/GT^2$

Step 2. Substitute the given information with units into the equation to solve for the mass of Saturn:

$M_{sat} = 4\pi^2(1.86 \times 10^8 \text{ m})^3/(6.67 \times 10^{-11} \text{ N} \cdot \text{m}^2/\text{kg}^2)(8.14 \times 10^4 \text{ s})^2$

$M_{sat} = 5.75 \times 10^{26}$ kg

Step 3. The orbital velocity of Mimas can be obtained from

$$\mathbf{v}_{orbit} = \sqrt{\frac{[(6.67\times10)]^{-11}\,\mathrm{N}\cdot\dfrac{\mathrm{m}^2}{\mathrm{kg}^2}\big)\big(5.75\times10^{26}\,\mathrm{kg}\big)}{1.86\times10^8\,\mathrm{m}}}$$

$$\mathbf{v}_{orbit} = 1.44 \times 10^4 \ \mathrm{m/s}$$

Problem-Solving Strategies to Avoid Missteps

When solving problems involving gravitation it is important to remember that the force between two objects is mutual. That means the force the earth exerts on the moon is equal to the force the moon exerts on the earth (the source of the tides). Gravitation is also an inverse square law, so do not forget to square the distance (center-to-center).

As always, consult the problem-solving ring to help you to organize your solution paths. Include correct units in all calculations. The gravitational field strength is equivalent to the acceleration due to gravity (**g**), and this value depends on the mass of the gravitating body and its radius. Newton's solution to the Kepler problem, in which he proved that closed orbits must be elliptical (or circular under special conditions), was a culminating triumph for classical mechanics in the seventeenth century.

Exercise 10.1

1. Calculate the gravitational force between a 2.5×10^{10}-kg mass and a 4.5×10^{12}-kg mass when the distance between them is 3.0×10^4 m. Assume that they can be treated as point masses.

2. Given the mass of the earth (5.98×10^{24} kg) and the radius of the earth (6.37×10^6 m), calculate the orbital velocity of a satellite that is 500 km above the surface of the earth.

3. Calculate the value of **g** on the surface of an imaginary planet that has a mass of 3.7×10^{23} kg and a radius of 5.5×10^5 m. How much would a 10-kg mass weigh on the surface of that planet?

4. Io is a satellite of the planet Jupiter. The orbital period is observed to be 1.77 days and its mean orbital distance from Jupiter (assume center-to-center) is 422,000 km. From this information, calculate the mass of Jupiter.

11

Fluids

In this chapter you will learn about why a boat floats and an airplane flies. These are not only practical concepts but very important applications of Newton's Laws. Previously, we applied Newton's Laws to solids; now we will apply them to fluids.

Static Pressure

Unlike solids, fluids do not have a definite shape and need to be contained. Liquids are described as *incompressible fluids*, and gases are described as *compressible fluids*. Regardless of the state of matter (solid, liquid, or gas), the force of gravity acts on the molecules of the substance. We define *pressure* as the force per unit area:

$$\mathbf{P} = \mathbf{F}/A \text{ (units are N/m}^2 = \text{pascals} = \text{Pa)}$$

In the case of a solid, the force is just the weight of the object and the area is measured by the surface area of contact. Therefore, two equal masses can exert different pressures if they have different surface areas. We see this in Figure 11.1 for a set of imaginary 2-kg blocks. Let us suppose that the contact surface area of block A is 4 m² and the surface contact area of block B is 8 m². The weight of each block is 19.6 N. Then

$$\mathbf{P}_A = (19.6 \text{ N})/(4 \text{ m}^2) = 4.9 \text{ N/m}^2$$
$$\mathbf{P}_B = (19.6 \text{ N})/(8 \text{ m}^2) = 2.45 \text{ N/m}^2$$

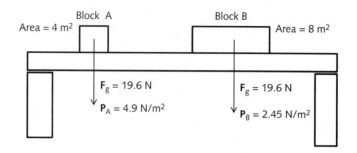

Figure 11.1 Force per unit area in a solid

For liquids (and gases), we imagine a cylinder with cross-sectional area A (Figure 11.2) filled to a depth d. The liquid has a density given by ρ = mass/volume (units of kg/m^3).

The density of water is defined to be ρ_{water} = 1 g/cm^3 = 1000 kg/m^3. Sometimes it is easier to use density units of g/cm^3, so we will convert back and forth as needed.

The force acting on the bottom of the cylinder is due to the weight of the fluid above it. Since $\rho = M/V$, this means that $M = \rho V$ and the weight of the fluid is $\mathbf{F}_g = mg = \rho g V$. The pressure at the bottom of the cylinder (due to the fluid) is now given by:

$$\mathbf{P} = \mathbf{F}/A = \rho g d \text{ (since volume/area = depth)}$$

Standard air pressure is usually taken to be (at sea level) equal to 1.01×10^5 N/m^2 and is also referred to as one atmosphere (1 atm).

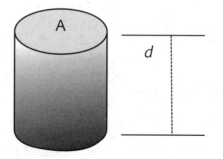

Figure 11.2 Measuring force per unit area for liquids and gases in a simple cylindrical container

Problem A liquid with a density of 13,600 kg/m³ (13.6 g/cm³) is poured inside a graduated cylinder to a depth of 0.76 m. Calculate the pressure at the bottom of the cylinder.

Solution

Step 1. We simply need to find the pressure $\mathbf{P} = \rho g d$ and substitute our given values.

Step 2. $\mathbf{P} = (13{,}600 \text{ kg/m}^3)(9.8 \text{ m/s}^2)(0.76 \text{ m}) = 1.01 \times 10^5 \text{ N/m}^2$

Important Tip

The liquid used in this problem is mercury and the level of 0.76 m (76 cm) is used in a barometer to measure air pressure (since the solution to the problem is 1 atmosphere). Also, notice that the pressure depends only on the depth of the liquid. If we used two cylinders (one wider than the other), then the pressures would be identical if the depths (and liquids) were identical!

Archimedes' Principle

Archimedes was an ancient Greek philosopher who demonstrated what happened when a solid is submerged in a liquid (typically water). There are two aspects to Archimedes' Principle:

1. A submerged object will always displace a volume of water equal to its own volume.

2. A submerged object will weigh less than it does in air by an amount equal to the weight of water displaced.

Experimentally, this is a useful concept. If you have an irregularly shaped object, you can determine its volume by submerging it in water and measuring the rise in the level of water. It is useful to recall that for liquid volume, 1 milliliter (1 mL) is approximately equal to 1 cubic centimeter (cm³). With regard to the apparent loss of weight in water, the upward displacement of the water (due to the submerged object) produces an upward force called the *buoyant force* ($\mathbf{F_B}$). Therefore, combining these ideas we see that

$$\text{Apparent weight in water} = \mathbf{F}_g - \mathbf{F}_B$$

Problem A 1.5-kg mass is fully submerged in water. It is observed that it weighs only 8.5 N when submerged. What is the volume of the mass?

Solution

Step 1. To find the volume of the mass we need to work backward from the buoyant force. If we know the buoyant force, we know the weight of the water displaced. If we know the weight of the water displaced we know the mass of the water displaced. If we express that mass in grams, and use the fact that the density of water is 1 g/cm³, we can then obtain the volume of the water displaced in cm³. However, using Archimedes' Principle, the volume of the water displaced is equal to the volume of the mass. This will solve the problem.

Step 2. First find the standard weight: $\mathbf{F}_g = m\mathbf{g} = (1.5 \text{ kg})(9.8 \text{ m/s}^2) = 14.7 \text{ N}$.

Step 3. The buoyant force = standard weight − apparent weight in water.

$\mathbf{F}_B = 14.7 \text{ N} − 8.5 \text{ N} = 6.2 \text{ N}$ (weight of the water displaced)

Step 4. Find the mass of the water displaced:

$M_{water} = (6.2 \text{ N})/(9.8 \text{ m/s}^2) = 0.63 \text{ kg} = 630 \text{ g}$

Step 5. Find the volume of the water displaced (which is equal to the volume of the mass).

630 g → 630 cm³ of water = 630 cm³ (volume of mass)

Problem   A 3000-g mass has a volume of 500 cm³. What is the apparent weight of the mass when it is fully submerged in water?

Solution

Step 1. We follow a similar procedure to solve the goal of finding the apparent weight of the mass when fully submerged in water. We know the volume of the object, and it will displace a volume of water equal to its own volume. The density of water is 1 g/cm³, so we immediately know the mass of the displaced water in units of grams. A quick conversion to kilograms allows us to obtain the weight of the displaced water (which is equal to the buoyant force). We can then determine the apparent weight in water. First, we know the standard weight:

$\mathbf{F}_g = m\mathbf{g} = (3 \text{ kg})(9.8 \text{ m/s}^2) = 29.4 \text{ N}$

Step 2. Volume of object = volume of displaced water = 500 cm³

Step 3. Mass of water displaced = 500 g = 0.5 kg

Step 4. Weight of water displaced = $\mathbf{F}_B = (0.5 \text{ kg})(9.8 \text{ m/s}^2) = 4.9 \text{ N}$

Step 5. Apparent weight in water = 29.4 N − 4.9 N = 24.5 N

Buoyancy

A submerged object has an upward force acting on it. This tendency to be pushed upward is called *buoyancy*. When an object is floating, then equilibrium is achieved and the buoyant force is equal to the weight (the net force is equal to zero). In the earlier example, the object would sink because the buoyant force is less than the weight of the object in air.

If an object is fully submerged in water, then it will be pushed upward if the buoyant force is greater than its standard weight and it will sink if the standard weight is greater than the buoyant force. If, while fully submerged, the buoyant force is equal to the standard weight, we say the object is neutrally buoyant and it will remain in place.

Large steel ships will float (even though the density of steel is greater than the density of water) if they displace a large enough volume of water. The density does play a role in the rule of flotation. If an object has a density that is half the density of water, then it will float with half of its volume submerged. An iceberg has approximately 90% of its volume under the water because its density is about 90% that of seawater (saltwater has a different density than freshwater).

Pascal's Principle

Because liquids are incompressible, they can do work against a piston if they are confined. This work is the main idea behind a hydraulic lift. French scientist/philosopher Blaise Pascal (1623–1662) developed the following concept: *any change in pressure applied at any given point of a confined fluid is transmitted undiminished throughout the fluid.*

We can express Pascal's Principle in terms of forces. If you recall that static pressure is defined to be equal to the force per unit area, then Pascal's Principle implies:

$$\mathbf{F}_1/A_1 = \mathbf{F}_2/A_2$$

A simple hydraulic system, sketched in Figure 11.3, operates like a simple machine. A small force applied to a small area transmits pressure undiminished through the contained liquid to a larger area that produces a larger force (to keep the ratio constant). With movable pistons, the small force may push a small piston down a large distance; the large-area piston on the other side rises a smaller distance but can lift a larger force (or lift a large weight).

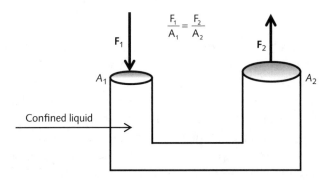

Figure 11.3 Simple hydraulic system illustrating Pascal's Principle

Problem A hydraulic system involves a 10-N force acting on the surface of a piston that has surface area of 5 m². If the piston on the other side has an area of 12 m², how much force can be applied? Assume this is a closed system.

Solution

Step 1. The goal of the problem is to find the lifting force in a hydraulic system using Pascal's Principle. We need to substitute the given information and solve for the second (lift) force.

Step 2. $(10 \text{ N})/(5 \text{ m}^2) = \mathbf{F}_2/(12 \text{ m}^2)$

$\mathbf{F}_2 = 24 \text{ N}$

Bernoulli's Principle

Even though we will not be treating the concepts of moving fluids mathematically, there are still some important concepts to consider. So far, we have discussed static fluids and static pressure. In 1738, Swiss mathematician Daniel Bernoulli (1700–1782) published a paper entitled *Hydrodynamica*, which dealt with the forces (dynamics) of fluids in motion (both liquids and gases). Fluids can be described by their *viscosity* (tendency to resist flowing). Expressing Bernoulli's statement in a simplified way: *for an ideal fluid (with little or no viscosity), an increase in the velocity of the fluid corresponds to a decrease in its static pressure.*

Bernoulli's Principle can be observed in the lift of an airplane wing. Figure 11.4 is a simple sketch of an airplane wing. The upper surface is

Figure 11.4 Airplane wing illustrating Bernoulli's Principle

curved or cambered while the bottom surface is essentially flat. As the air flows over the upper surface, it speeds up and the static pressure on the upper surface decreases relative to that of the bottom surface. This pressure difference results in lift and is one of the many reasons why an airplane flies. You can see Bernoulli's Principle in action for yourself by blowing across a piece of paper and watching it rise.

Problem-Solving Strategies to Avoid Missteps

As with all physics problems, consult the problem-solving ring if you need assistance. It is important to remember that the buoyant force is not equal to the *volume* of the liquid displaced (typically, water), but to the *weight* of the liquid displaced. Additionally, the density of water is defined as 1 g/cm³ (and 1 mL is approximately equal to 1 cm³), so it may be easier to do simple buoyancy problems with masses in grams.

The buoyant force may be more than, less, or the same as the weight of an object (when the object is submerged), but when floating, the object is in equilibrium and the buoyant force is equal to the weight. Remember, it is the volume of the object that displaces a large enough amount of water to float. In the same way, the lifting force on an airplane is due to the *difference* in pressure between the upper and lower surfaces of the wing.

Exercise 11.1

1. What would be the length of a column of water needed to balance 1 atmosphere of air pressure (1.01×10^5 N/m²)?

2. A 200-g mass has a volume of 500 cm³. It is fully submerged in water.

    a. What will be the mass of the water displaced in grams and kilograms?

    b. What will be the magnitude of the buoyant force?

    c. Will the object float based on your answer to part (b)?

3. A 200-N mass is fully submerged in water. When it is fully submerged it is observed to weigh only 50 N.

   a. What is the magnitude of the buoyant force?

   b. What is the volume of the object?

4. A hydraulic lift is used to apply a force of 1000 N over a piston with an area of 30 m². If a force of only 50 N is desired, how much area should the piston have?

12

Heat and the Kinetic Theory of Gases

In this chapter you will learn about *internal energy*. This is more commonly called *heat*. On the molecular level, we take a brief look at the kinetic theory of gases. This is a detour away from the study of large masses and their inter-actions. Because inelastic collisions involve the loss of kinetic energy and because friction also removes kinetic energy from a system, a brief study of the background concepts seems reasonable.

Temperature Measurements

It is easy to observe that friction transfers heat (or internal energy). Rub your hands together on a cold day and place them against your cheeks. Do you feel their warmth? If you place a box painted black outside under the sun for a long time, it will get hot. It may even expand in size, but it will not weigh more. What is going on?

Before we study the molecular model of heat, we can first make objec-tive measurements. The effects of heat transfer can be measured using a thermometer. In physics, either of two measurements of temperature is used. Typically, we use the *Celsius* scale (sometimes known as *centigrade* because of its metric nature), named for Swedish scientist Anders Celsius (1701–1744). At standard air pressure, pure water solidifies (freezes) at 0°C and the same water vaporizes (boils) at 100°C.

The other system of measurement (usually represented without the degree symbol) is called the *Kelvin* scale, named after the British scientist Lord Kelvin (William Thomson, 1824–1907). In the Kelvin scale, a factor of 273 is added to the Celsius temperature (thus, water freezes at 273 K and boils at 373 K under standard atmospheric conditions). Part of the reason for developing this scale was the recognition by physicists that at approximately –273°C, molecular motion reaches a minimum. This temperature is also known as *absolute zero* (0 K). In physics we do *not* use the Fahrenheit scale (named for Daniel Fahrenheit, 1686–1736).

Misconception

Temperature and heat are *not* the same thing. Your teacher may have made this quick demonstration: If equal amounts of water are placed in two separate beakers (it would be better to use an *insulating* material, which limits the amount of heat conduction between materials), at equilibrium, they should measure the same temperature. If unequal amounts of hot water are poured in (unequal amounts but the same temperature) and the temperature recorded, you will see that different amounts of heat are transferred between the hot and cold water (proportional to the mass or volume added) even though the hot water added was initially the same temperature. *Warning: It can be dangerous to do experiments with heat without proper equipment and supervision!*

Problem Convert the Celsius temperatures to Kelvin units and the Kelvin temperatures to Celsius units:

a. 25°C = _____ K

b. –175°C = _____ K

c. 2500°C = _____ K

d. 5000 K = _____°C

e. 30 K = _____°C

f. 175 K = _____°C

Solution

Step 1. We simply add 273 to all the Celsius temperatures and subtract 273 from all the Kelvin temperatures.

Step 2. The measures are:

a. 25°C = 298 K

b. –175°C = 98 K

c. 2500°C = 2773 K

d. 5000 K = 4727°C

e. 30 K = –243°C

f. 175 K = –98°C

Heat Transfer

Heat is a measure of thermal energy. The heat content (Q), in joules, of a material is determined by the type of material, its mass, and its temperature. We define the specific heat capacity of a material as the amount of energy (ΔQ in joules) needed to raise the temperature of 1 gram of a substance by 1 degree Celsius (in units of joules per gram per degree Celsius or kilojoules per kilogram per degree Celsius. We use the lowercase letter c to represent specific heat:

$$c = \Delta Q / m\Delta T$$

The specific heat of water is defined as approximately 4.19 J/g·°C or 4.19 kJ/kg·°C. The values of different specific heats are listed in Table 12.1.

We can rearrange the definition of specific heat to find an equation for heat transfer in a given material (in a given phase or state):

$$\Delta Q = mc\Delta T$$

Table 12.1   Specific Heats of Various Substances

SUBSTANCE	SPECIFIC HEAT (J/G·°C)
Aluminum	0.900
Bismuth	0.123
Copper	0.386
Brass	0.380
Gold	0.126
Lead	0.128
Silver	0.233
Tungsten	0.134
Zinc	0.387
Mercury	0.140
Alcohol(ethyl)	2.4
Water	4.19
Steam	2.01
Ice	2.05
Granite	0.790
Glass	0.84

Problem   How much energy is transferred when 50 g of brass at 20°C is heated to a temperature of 150°C?

Solution

Step 1. The goal is to find the amount of heat transferred in joules (ΔQ). Brass has a particular specific heat and so we need to locate that on the chart (Table 12.1). We are assuming that the brass is not melting (see the next section, which discusses energy transfers that accompany a change of state).

Step 2. Substitute in the given values with units:

$$\Delta Q = mc\Delta T = (50 \text{ g})(0.380 \text{ J/g} \cdot °C)(130°C) = 2470.0 \text{ J}$$

Problem   In an experiment, a student wishes to determine the specific heat of an unknown substance. It is observed that 100 g of the substance absorbs 3000 J of energy as it is heated from 30°C to 250°C with no change of state. What is the value of the specific heat of this material?

Solution

Step 1. To find the specific heat of the material we use our definition equation:

$$c = \Delta Q / m\Delta T$$

Step 2. Substitute in the given values with units:

$$c = (3000 \text{ J})/(100 \text{ g})(220°C) = 0.136 \text{ J/g} \cdot °C \text{ (very close to tungsten)}$$

Heat transfer can take place between two materials. Assuming no gain or loss of heat to the surrounding space, then thermal energy is conserved:

$$\text{Heat gain} = \text{Heat loss}$$

or

$$\Delta Q_{gain} = \Delta Q_{loss}$$

and

$$m_1 c \Delta T_{gain} = m_2 c \Delta T_{loss}$$

where

$$\Delta T = T_f - T_i$$

Problem 100 g of water at 20°C is mixed with 150 g of water at 85°C. Assuming no loss to the surrounding space, what is the value of the final temperature of the mixture?

Solution

Step 1. To find the final temperature of the mixture, we know that heat gain equals heat loss: $\Delta Q_{gain} = \Delta Q_{loss}$

Step 2. To keep the magnitude of the gain and the loss positive, we can write:

$$m_{cw}c_{cw}(T_f - T_i) = m_{hw}c_{hw}(T_i - T_f)$$

Step 3. Substitute in the given values with units and solve for the final temperature. Note that the specific heat of water (hot or cold) is irrelevant to the problem since we are mixing the same substance (the specific heats are canceled out):

$$(100 \text{ g})(T_f - 20°C) = (150 \text{ g})(85°C - T_f)$$

and

$$T_f = 59°C$$

Problem 50 g of zinc at 100°C is placed into 100 g of water at 20°C. What is the final temperature of the mixture (assume no loss of heat to surrounding space)?

Solution

Step 1. We cannot cancel out the specific heats because we have two different substances. We are still assuming that heat is conserved and there is no loss of heat to the surrounding space:

$$\Delta Q_{gain} = \Delta Q_{loss}$$

Step 2. We write our equations as follows for the zinc and cold water:

$$m_{zinc}c_{zinc}\Delta T_{zinc} = m_{cw}c_{cw}\Delta T_{cw}$$

Step 3. $(50 \text{ g})(0.387 \text{ J/g}\cdot°C)(100°C - T_f) = (100 \text{ g})(4.19 \text{ J/g}\cdot°C)(T_f - 20°C)$

$$T_f = 23.5°C$$

Even though we will not discuss the following concepts in detail, it is worth noting that heat can be transferred in three ways: conduction, convection, and radiation. Conduction involves molecular collisions. Convection involves the large-scale movement of material (for example, warming a room by circulating rising hot air and sinking cold air). Radiation involves the transfer of energy using waves (see Chapter 13).

Change of State

At normal atmospheric pressure, water boils and changes state from a liquid to a gas at 100°C. The temperature, however, does not change, as the energy is used to change the state of the matter (solid, liquid, or gas). The energy needed to change the state of matter from liquid to gas (or gas to liquid) is called the *latent heat of vaporization* (H_v in units of J/g or kJ/kg). The energy needed to change the state of matter from solid to liquid (or liquid to solid) is called *the latent heat of fusion* (H_f in units of J/g or kJ/kg).

$$\Delta Q = mH_f \quad \text{and} \quad \Delta Q = mH_v$$

Table 12.2 lists the boiling points, melting points, heat of fusion, and heat of vaporization for various substances. If a substance was heating up and then changing states, you would need to add all of the thermal energy changes involved to determine the total amount of energy transferred. Refer to Tables 12.1 and 12.2 as needed for specific heats and the respective heats of fusion and vaporization.

Problem How much energy is required to change 50 g of lead from a solid at 100°C to a liquid at 328°C?

Solution

Step 1. To solve this problem, we recognize that there are two changes taking place. First, the lead heats up as a solid from 100°C to its melting point at 328°C. Secondly, at its melting point, the lead liquefies.

Step 2. We now calculate the heat changes for each process.

Lead as a solid: $\Delta Q = mc\Delta T = (50 \text{ g})(0.128 \text{ J/g} \cdot \text{C})(228°\text{C}) = 1459.2 \text{ J}$

Lead in melted form: $\Delta Q = mH_f = (50 \text{ g})(25 \text{ J/g}) = 1250 \text{ J}$

Step 3. We add the total energy changes to get $\Delta Q_{\text{total}} = 2709.2 \text{ J}$

Table 12.2 Temperatures for Changes of State for Various Substances

SUBSTANCE	MELTING POINT (°C)	BOILING POINT (°C)	HEAT OF FUSION (J/G)	HEAT OF VAPORIZATION (J/G)
Alcohol(ethyl)	–117	79	109	855
Aluminum	660	2467	396	10,500
Ammonia	–78	–33	332	1370
Copper	1083	2567	205	4790
Iron	1535	2750	267	6290
Lead	328	1740	25	866
Mercury	–39	357	11	295
Platinum	1772	3827	101	229
Silver	962	2212	105	2370
Tungsten	3410	5660	192	4350
Water (ice)	0	—	334	—
Water (liquid)	—	100	—	2260
Zinc	420	907	113	1770

Misconceptions

Some of you may recall that heat may be measured in units of calories. These are old units no longer used in physics. In this context, the calorie is defined as the amount of heat needed to raise the temperature of 1 gram of water by 1 degree Celsius. James Joule was the first to develop the concept of the *mechanical equivalent of heat* and experimentally demonstrated that in water, 1 calorie is equivalent to 4.19 joules. Thus, the specific heat of water would be defined as 1 cal/g·°C. We use SI units, so the specific heat of water is listed on our reference tables as 4.19 J/g·°C (4.19 kJ/kg·°C). Do not confuse *food calories* with heat calories. Actually, 1 *food* calorie is equivalent to 1 *kilocalorie* (kcal) of heat.

When studying heat transfers in a substance, sometimes a graph of energy vs. temperature is used. Figure 12.1 is a graph of energy vs. temperature for 20 g of an imaginary substance that begins as a solid at 0°C.

Problem For the graph shown in Figure 12.1, find:

a. The melting point

b. The boiling point

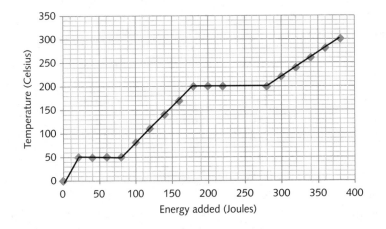

Figure 12.1 Graph of energy vs. temperature

    c. The specific heat as a solid

    d. The specific heat as a liquid

    e. The specific heat as a gas

    f. The heat of fusion

    g. The heat of vaporization

Solution

Step 1. The graph can be analyzed for many of the key concepts just discussed. If we know that the substance begins in the solid state at
0°C, then (a) the melting point is seen as 50°C (this is where the
temperature is not changing as the material changes from a solid
to a liquid) and (b) the boiling point is observed to be equal to
200°C.

Step 2. The specific heat of the substance as (c) a solid can be obtained using
the fact that 20 joules of heat are used to raise the temperature of
the material from 0°C to 50°C. Hence

$$c(\text{solid}) = \Delta Q/m\Delta T = (20 \text{ J})/(20 \text{ g})(50°C) = 0.02 \text{ J/g} \cdot °C$$

The specific heat of the substance as (d) a liquid and (e) a gas can be
found by a similar analysis:

$$c(\text{liquid}) = (100 \text{ J})/(20 \text{ g})(150°C) = 0.033 \text{ J/g} \cdot °C$$

The specific heat of the substance as a gas can also be found:

$c(gas) = (100 \text{ J})/(20 \text{ g})(100°C) = 0.05 \text{ J/g} \cdot °C$

Step 3. (f) The heat of fusion can be found by looking at the first horizontal segment of the graph:

$H_f = \Delta Q/m = (60 \text{ J})/(20 \text{ g}) = 3 \text{ J/g}$

(g) The heat of vaporization can be found using the second horizontal segment of the graph:

$H_v = \Delta Q/m = (100 \text{ J})/(20 \text{ g}) = 5 \text{ J/g}$

Kinetic Theory of Gases

Gases, we have learned, are compressible fluids. Gases are made of molecules, and it is important to consider the effects of molecular collisions inside of a gas and what role they play in the generation of what we perceive as *heat*.

The *kinetic theory of gases* states (in a simplified form) that we can consider the collisions of molecules in an ideal confined gas to be elastic. The average molecular kinetic energy is observed by us as the absolute (Kelvin) temperature, and the transfer of momentum, due to collisions with the walls of the container, is observed by us as the pressure.

While we will not go into deep detail in this book, the relationship between volume, pressure, and absolute temperature in an ideal gas can be expressed in the ideal gas law (where the temperature must be in kelvins):

$$\frac{P_1 V_1}{T_1} = \frac{P_2 V_2}{T_2}$$

At constant temperature, the ideal gas law (in this form) reduces to *Boyle's Law*:

$$P_1 V_1 = P_2 V_2$$

At constant pressure, the ideal gas law reduces to Charles's Law:

$$\frac{V_1}{T_1} = \frac{V_2}{T_2}$$

Misconceptions

The kinetic theory of gases involves much more than just the three equations shown earlier. The motive of this book is to provide the reader with an introduction to the material and not a detailed derivation or review of all material related to any subject. More details on the kinetic theory of gases and the so-called p-V diagrams can be found in any introductory textbook. Notice that the units for PV [(N/m²) (m³)] are joules!

Additionally, it is important to remember that although the equations are defined only for Kelvin temperatures, the ratio comparisons imply that the pressure can be expressed in atmospheres and the volume can be expressed in liters (instead of N/m² or m³, respectively).

Problem An ideal gas at 50°C has a volume of 100 m³ at a pressure of 1.5 atm. If the temperature is increased to 100°C, and the volume is reduced to 80 m³, calculate the new pressure of the gas.

Solution

Step 1. We are given the pressure, volume, and temperature, so we can use the ideal gas equations to solve for the new pressure. However, the temperature must be converted to kelvins.

Step 2. Note that although the Celsius temperature is doubled, the Kelvin temperature is *not* doubled! The temperature conversions are:

$T_1 = 50°C = 323 \text{ K}$

and

$T_2 = 100°C = 373 \text{ K}$

Step 3. Substitute in the given values with units and solve for the new pressure:

$(1.5 \text{ atm})(100 \text{ m}^3)/(323 \text{ K}) = P_2(80 \text{ m}^3)/(373 \text{ K})$

$P_2 = 2.2 \text{ atm}$

Problem-Solving Strategies to Avoid Missteps

When solving problems concerning heat and the gas laws, it is important to remember the proper units. For the gas laws, the temperature must be in kelvins. For heat transfer problems, the units for specific heat are in units of J/g·°C or kJ/kg·°C. To convert from the Celsius to the Kelvin scale, you add 273. Absolute zero (0 K) is equal to –273°C and is the lowest possible temperature.

When interpreting heating curves, it is important to know how to find the specific heats in the different states of matter using the data given on the graph. Typically, the temperature is plotted on the vertical axis, so the specific heat is *not* equal to the slope of the line since

$$c = \Delta Q/m\Delta T$$

Exercise 12.1

1. Calculate the final temperature of the mixture when 100 g of aluminum at 200°C is dropped into 400 g of water at 20°C (assume no loss of heat to the surroundings).

2. A heating curve for 30 g of an imaginary substance is shown (it starts as a solid at 0°C).

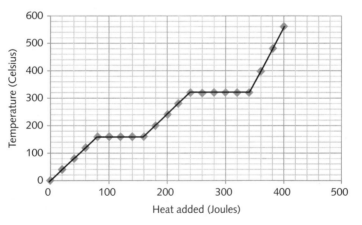

Find:

a. The specific heat as a solid, liquid, and gas.

b. The heats of fusion and vaporization.

3. An ideal gas is confined under a pressure of 2 atm and a temperature of 75°C. The initial volume is 500 m³. If the volume is halved and the pressure tripled, calculate the new temperature of the gas.

13

Waves

In this chapter you will learn about the properties of mechanical waves and sound waves. These include waves in springs, strings, water, and air. Building a mechanical model of waves and demonstrating how they transport energy is an important step in our journey of exploration. The study of mechanical waves will be important later on for our study of light and optics.

Properties of Mechanical Waves

Waves are periodic disturbances in a medium. When an elastic material oscillates up and down (or back and forth), the disturbance travels in the form of a wave. We see waves all around us. Sound travels through the air in the form of waves. If you throw a stone into a pond, you will see circular ripples moving out from the center. A stretched spring can display waves. And who is not moved by a piece of beautiful music? Later on, we will show that even light demonstrates properties of waves. By working with mechanical oscillations, it is possible to build up properties that are characteristic of all waves.

First, we introduce some important terms. A *pulse* is a single vibratory disturbance in an elastic medium. If the direction of the vibration is *perpendicular* (imagine a string moved up and down while tied at both ends) to the apparent direction of motion, then the pulse is called *transverse*. If the direction of the vibration is *parallel* to the apparent direction of motion, then the pulse is called *longitudinal*. It is very important to remember that the pulse (and waves in general) transmits only energy (see Figure 13.1). The height of the pulse is called its *amplitude*, and the velocity of the pulse depends on the properties of the medium (for example, its density, tension, and inertial properties).

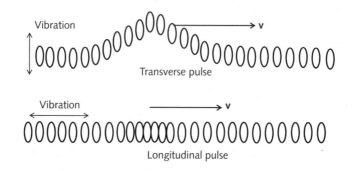

Figure 13.1 Wave pulses

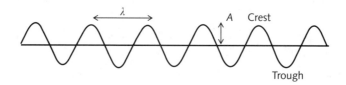

Figure 13.2 Transverse wave

A wave is described as a continuous set of pulses making complete oscillations within a definite period of time. If points in the medium have the same motion, simultaneously, then those points are said to be *in phase*. In a transverse wave, the height above the rest or equilibrium position corresponds to the amplitude (*A*) of the wave. The distance between any two successive points in phase is called the *wavelength* (designated by the Greek letter λ, lambda). The number of complete cycles (or waves per second) is called the *frequency* (*f*), measured in units of cycles per second (s⁻¹), or hertz (Hz). The time to complete one wave (or one cycle) is called the period (*T*) and it is equal to the reciprocal of the frequency (*T* = 1/*f*). Figure 13.2 illustrates these ideas in a transverse wave. In a transverse wave the high and low points are typically referred to as *crests* and *troughs*, respectively.

For a longitudinal wave, we observe periodic regions of compressions and expansions. These expansions are sometimes called *rarefactions*. Longitudinal waves can likewise be characterized by amplitudes, wavelengths, frequencies, and periods.

The velocity of a wave is given by the product of its frequency times the wavelength:

$$\mathbf{v} = f\lambda$$

The velocity will be constant if the conditions of the medium remain constant. Therefore, the wavelength and frequency will be inversely proportional to each other.

Misconceptions

It is important to remember that waves transfer only energy. The particles in the elastic medium are either moving up and down (transverse wave) or back and forth (longitudinal wave). In a transverse wave, there are actually many different ways (actually an infinite number) to have vibrations perpendicular to a chosen direction of motion. The ability to select preferred orientations for vibrations is a characteristic of transverse waves known as *polarization*. Longitudinal waves *cannot* be polarized!

Problem   A wave is formed by moving a string up and down. Crests are measured every 0.3 m and a period of 0.25 s is observed. Calculate the frequency and velocity of the wave.

Solution

Step 1. The goal of this problem is to calculate the velocity of the wave. To accomplish this, we first determine the frequency of the wave:

$$f = 1/T = 1/(0.25 \text{ s}) = 4 \text{ Hz}$$

Step 2. To calculate the velocity, we use the equation:

$$\mathbf{v} = f\lambda = (4 \text{ Hz})(0.3 \text{ m}) = 1.2 \text{ m/s (recall that 1 Hz = 1 s}^{-1} \text{ in SI units)}$$

It can be shown that the velocity of the wave increases with the tension in a string (or spring). The reflection of a one-dimensional transverse pulse can be observed when a string is tied at one end. It will be observed (see Figure 13.3) that the pulse returns *inverted* from its original phase orientation.

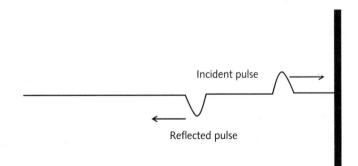

Figure 13.3 Reflection of a transverse pulse

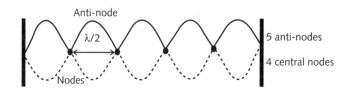

Figure 13.4 Standing wave

If transverse waves are sent along a string tied at both ends, and if the frequency is timed correctly, then a standing wave occurs as the incoming waves mix (superpose) with the inverted reflected waves. The resultant wave will appear to be standing in one place (only showing an up-and-down motion). Points along the wave where there is no appreciable motion are called *nodes* (the maximum displacement points are called *anti-nodes*). Nodes appear every half-wavelength in a standing wave (see Figure 13.4).

Notice that if the frequency is increased (or decreased), the number of nodes would increase (or decrease).

Problem A spring is stretched to a distance of 5.0 m and held under constant tension. A transverse pulse is generated and takes 1.8 s to travel from one end to the other, and then back again.

a. What is the velocity of the pulse?

b. Standing waves are now generated and nodes are observed every 25 cm. What is the wavelength of the wave?

c. Based on your answers to parts (a) and (b), determine the frequency of the wave (assume the tension stays constant).

Solution

Step 1. (a) The velocity of the pulse is simply the distance divided by the time to go one length down the spring (we are given the round-trip time!).

$\mathbf{v} = d/t = (5.0 \text{ m})/(0.9 \text{ m}) = 5.55 \text{ m/s}$

Step 2. Because the conditions of the medium are constant (constant tension), we can produce standing waves. The nodes appear every half wavelength. (b) It is observed that the nodes appear every 25 cm, hence:

$\lambda = 50 \text{ cm} = 0.50 \text{ m}$

Step 3. To find (c) the frequency, we note that the velocity of any wave in this spring will be the same since the conditions are constant. This means that we can use the equation:

$\mathbf{v} = f\lambda$

$5.55 \text{ m/s} = f(0.50 \text{ m})$

$f = 11.1 \text{ Hz}$

This concept of mixing (called *superposition*) waves leads to the concept of *interference*. If we have one-dimensional waves in a stretched string (or spring), then two pulses (let us say oriented upward with amplitudes *a* and *b*, respectively) moving toward each other will *constructively interfere* when they meet. That is, at the point that they meet the material will be pushed upward and produce a resultant pulse with an amplitude equal to *a + b*. The pulses will then continue onward undiminished (see Figure 13.5).

If the phase orientation on pulse *a* is reversed (upside-down), then when the pulses meet at point *p*, they *interfere destructively* (*a − b*) before continuing onward (see Figure 13.6). Assuming equal amplitudes (opposite phases), then *a − b* = 0.

If we generate two-dimensional waves (let's say in a water tank), then we can see more wave properties. A point source will generate *wave fronts* that

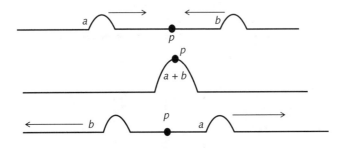

Figure 13.5 Constructive interference

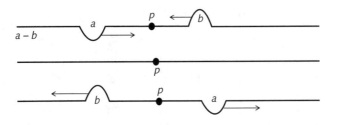

Figure 13.6 Destructive interference

are circular. If we have a straight-wave generator, then the wave fronts will be known as *plane parallel* or just *straight* wave fronts. If plane waves are incident on a barrier at some angle (measured to a normal line drawn perpendicular to the barrier), then the waves will be reflected back at an angle equal to the angle of incidence. This is known as the Law of Reflection (see Figure 13.7).

If the plane waves are incident on the edge of a barrier or on a small opening (aperture), then the wave will diffract around the barrier (see Figure 13.8). The explanation of this wave behavior is due to Christian Huygens (1629–1695), who stated that when plane waves are incident on a small opening, the opening acts like a point source and produces circular (or, in three dimensions, spherical) wave fronts. It is important to remember that interference and diffraction are phenomena only characteristic of waves (*not* particles)!

The interference of two-dimensional waves can be demonstrated in a water (ripple) tank and modeled using a series of concentric circles. If two point sources produce circular waves (see Figure 13.9), then where the crests (the outer edge of each circle) overlap, we have regions of constructive interference. Where crests and troughs overlap (midway), there will be

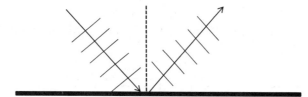

Figure 13.7 Reflection of plane waves

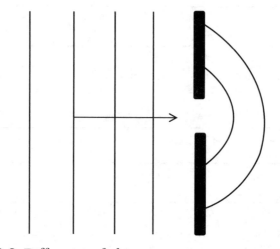

Figure 13.8 Diffraction of plane waves

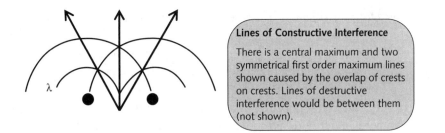

Lines of Constructive Interference

There is a central maximum and two symmetrical first order maximum lines shown caused by the overlap of crests on crests. Lines of destructive interference would be between them (not shown).

Figure 13.9 Point source interference

regions of destructive interference. We will examine this model in greater detail when we study the concepts of light.

Sound

Sound is a longitudinal mechanical wave. At 0°C, the velocity of sound (at sea level) is approximately 331 m/s. The velocity of sound increases by approximately 0.6 m/s, for each 1°C rise in temperature. The amplitude of the sound waves corresponds to its loudness. The frequency of the sound wave is commonly known as the *pitch*. The wave properties of sound can be demonstrated by showing that sound diffracts around the edges of walls and buildings, a phenomenon known as *beats*, which is a warbling (loud then soft) interference sound pattern heard when two tones with slightly different frequencies are played simultaneously. We previously stated that sound (created by pressure differences in the air) *cannot* be polarized and is therefore *not* a transverse wave.

Problem   A tuning fork is listed as having a vibrational frequency of 256 Hz. If the air temperature is 25°C, calculate the wavelength of the sound produced when the fork is struck.

Solution

Step 1. Before we can determine the wavelength, we need to determine the velocity of sound. At 25°C, the velocity of sound is

$\mathbf{v}$ = 331 m/s + (0.6 m/s)(25°C) = 346 m/s

Step 2. We can now use our equation for the velocity of a wave to calculate the wavelength:

$\mathbf{v} = f\lambda$

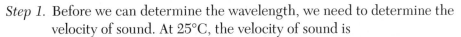

346 m/s = (256 Hz)λ and λ = 1.35 m

The *Doppler effect* is often associated with sound. If we have a stationary observer and a moving source of sound waves, then there will be an apparent increase in the frequency of the sound as the source moves toward the observer and an apparent decrease in the frequency of the sound as the source moves away from the observer.

If the source (generating a sound with frequency f and moving with velocity $\mathbf{v}_s$) is moving *toward* the stationary observer, then the observed frequency f' is given by:

$$f' = f\,[\mathbf{v}/(\mathbf{v} - \mathbf{v}_s)], \text{ where the wave velocity in the medium is } \mathbf{v}$$

If the source is moving *away* from the stationary observer, then the observed frequency is given by:

$$f' = f\,[\mathbf{v}/(\mathbf{v} + \mathbf{v}_s)]$$

Notice for a source moving *toward* a stationary observer that if the source velocity is *equal* to the wave velocity, the Doppler equation is *undefined* (divided by zero). This situation produces a *shock wave*, which would be heard as a *sonic boom*.

Problem   An ambulance has a velocity of 25 m/s and is producing a steady frequency from its siren of 256 Hz. What is the value of the apparent frequency heard by an observer who is in front of the ambulance? Assume a velocity of sound equal to 343 m/s.

Solution

Step 1. The problem is all set up to solve for the apparent frequency. All we need to do is to substitute in our given values with units:

$$f' = (256 \text{ Hz})\,[(343 \text{ m/s})/(343 \text{ m/s} - 25 \text{ m/s})] = 276 \text{ Hz}$$

Resonance is a phenomenon often associated with sound. Every object has a natural vibrating frequency. If a tuning fork is struck, it will send out sound waves. If an identical tuning fork is nearby, it too will vibrate (in resonance) because it has the same natural vibrating frequency. The word *resonance* itself comes from the Latin meaning "an echo" (to "re-sound"). Musical instruments (such as stringed or woodwind instruments) operate on the principle of a resonating sound box. The air inside the sound box resonates at the same frequency as the vibrating strings (for example) and produces the amplified sound. A child on a swing demonstrates resonance as she pumps the swing by stretching out her legs and then tucking them in at correct positions. In this way, she can build up a large amplitude with just a small initial push.

The concept of wave *refraction* will be studied later in Chapter 17. When a wave travels from one medium to another, some of the wave's energy is transmitted, some of the energy is reflected, and some of the energy is absorbed. If the wave changes direction as a result of changing velocity (at an oblique angle), then the direction of propagation will change and we say that the wave has been *refracted*.

Finally, we will learn that while mechanical waves require a medium for propagation, light, referred to as an electromagnetic wave, does not require a medium to propagate. To understand what an electromagnetic wave is, we must first understand the concepts of electricity and magnetism. We begin this study in the next chapter.

There are many other aspects of wave phenomena that we did not have space to cover in this chapter. This book is only a brief survey of some of the major topics covered in a typical high school physics class.

Problem-Solving Strategies to Avoid Missteps

It is important to remember all of the vocabulary associated with waves. The difference between transverse and longitudinal waves is very important. Waves transfer only energy, and understanding their properties is part of our journey of exploration in physics. The problem-solving ring will allow you to identify the best solution path for solving various problems concerning mechanical waves. As always, be very careful with units, and remember that the velocity of a wave will change as the properties of the medium change.

Exercise 13.1

1. A police car siren emits a frequency of 430 Hz while traveling at 45 m/s away from a stationary observer. If the air temperature is 27°C, calculate the apparent frequency heard by the observer.

2. A tuning fork produces waves in a hollow air tube at a frequency of 512 Hz. Standing waves are produced in a resonance tube and nodes are measured every 35 cm. Calculate the temperature of the air inside the tube.

3. Waves with a period of 0.05 s are produced in a stretched spring (held under constant tension). Calculate the wavelength of the waves if the observed wave velocity is 5 m/s.

14

Static Electricity

In this chapter you will learn about the properties of static electricity. We leave the macroscopic world of masses to begin exploring the microscopic world of electricity. Our modern world of electronics, cell phones, and computers has come a long way from Thomas Edison's invention of the light bulb and the humble experiments of Benjamin Franklin flying a kite in a thunderstorm. Our journey through physics continues over the next three chapters as we learn about electric circuits, magnetism, and electromagnetism.

Electric Charges

Anyone who has seen lightning during a thunderstorm can recognize the power of electric charges. Anyone who has gotten a *shock* after walking across a carpeted floor in their socks also recognizes the power of electric charges.

The study of static electricity has its recorded beginnings in ancient Greece. Like the Greek word *phyiska* (the study of natural things or natural philosophy), the rubbing of amber (in Greek, *elektron*) with silk produced the effect of lifting pieces of parchment. You can achieve the same effect by running a plastic comb through your hair (or rubbing it on a sweater).

The cause of this phenomenon was poorly understood. Benjamin Franklin demonstrated that lightning was electricity and introduced the terms *positive* and *negative electricity* to describe what he thought was an electric fluid. An easy demonstration you can do is with ordinary cellophane tape. Place two equal lengths of tape next to each other on a desk (sticky side down). Then pull them off the desk quickly. When you bring them near to each other, you will notice that they repel each other (evidence of a force; recall Newton's Laws). If, however, you place the two pieces of tape on top of each other (with the bottom one attached to the desk), then pull them off and separate them, the two pieces of tape will attract each other.

This property of matter is called *charge* and we now know that there exist two fundamental *quanta* of charge. We call these *elementary charges*, and ordinary matter is electrically neutral when it contains an equal number of positive and negative charges. Ions have either more positive or more negative charges. From atomic theory, we know that electrons are the fundamental carriers of *negative charge* and protons are the fundamental carriers of *positive charge*. In solids, only electrons are transferred.

Materials that do *not* allow electrons to easily move through them are called *insulators*. Materials that allow electrons to easily move through them are called *conductors*. Typical insulators would include glass, rubber, plastic, wood, and paper. All metals are conductors due to the fact that they have large numbers of free electrons in their atomic structure. Metals can be classified by their *conductivity* (ability to conduct electrons through them) or *resistivity* (ability to inhibit the flow of electrons through them). Copper (a typical metal used for wires) has a relatively high *conductivity* and a relatively low *resistivity*. Tungsten, a metal used in high-intensity light bulbs because of its high melting point, has a relatively high resistivity. We shall study the implications of resistivity and electric current in Chapter 15.

Charge is a fundamental quality of matter. It is never created or destroyed (conservation of charge). According to the electron model for charge transfer, objects become charged through the transfer of electrons (or elementary charges). We will not be studying ions or ionic solutions in this book (check any standard chemistry book for more information on electrolytic solutions and ionization of gases). In modern units, we assign a unit of charge called a *coulomb* (named after French scientist Charles Augustin de Coulomb, 1736–1806) when 6.25×10^{18} elementary charges have been transferred. The coulomb (C) is actually a unit *derived* from electric current. The unit for *electric current* (a *fundamental* SI unit) is called an *ampere* (A), named for French scientist André Marie Ampère (1775–1836). This corresponds to the flow of charge in a conductor. The coulomb is defined as equal to 1 ampere × second $(1 \text{ C} = 1 \text{ A} \cdot \text{s})$, and $1 \text{ A} = 1 \text{ C/s}$. This means that one electron has a charge of -1.6×10^{-19} C.

When charges are transferred, electrons will flow from a high concentration to a low concentration. In our model, an object becomes *positively* charged by losing electrons, and an object becomes *negatively* charged by gaining electrons (if one object loses 2 C of charge, another must gain 2 C of charge!). *Electrostatics* is the name we give to the study of charges at rest, or *static electricity*.

Certain materials like rubber or vinyl tend to become *negatively* charged, whereas other materials like glass or plastic tend to become *positively* charged. Like charges will repel and unlike charges will attract. Insulators can be charged by frictional contact (such as rubbing a comb through your hair).

Metals cannot be charged *statically*, but if they are placed on insulating materials, they can act like static charges and be charged by *contact, conduction,* or *induction.* Charging by *contact* would involve touching a metal object (like a sphere on an insulating stand; see Figure 14.1) with another charged object (such as a negatively charged rubber rod that had been rubbed with fur or wool). The sign of the charge would be the same on both objects.

Charging by *conduction* would involve using a conducting medium (like a wire or any other conducting material) to transfer the charges from one object to another (see Figure 14.2). Charging by *induction* does not involve

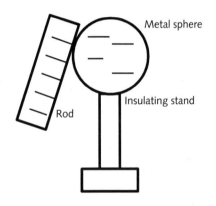

Figure 14.1 Charging by direct contact

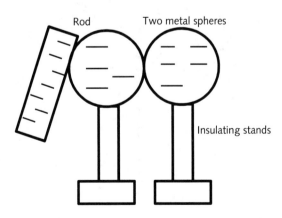

Figure 14.2 Charging two metals

any direct contact. One typical way to charge by induction is by using a ground (transferring charges to or from the earth by means of a conductor; a lightning rod is an example of a ground connection). Figure 14.3 shows a metal sphere being charged positively by induction using a negatively charged rod and a ground connection. A ground connection can be a wire directly connected to the ground or to a metal pipe.

If the sphere is initially neutral, and a charged rod is brought nearby, then the negative charges are repelled to the farthest side of the sphere and we have a condition known as *charge polarization* (see Figure 14.4). This

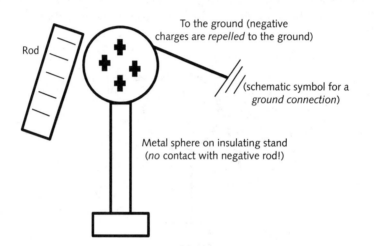

Figure 14.3 Charging a metal sphere positively by induction

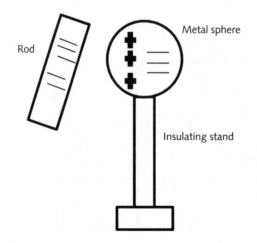

Figure 14.4 Charge polarization in a neutral metal sphere

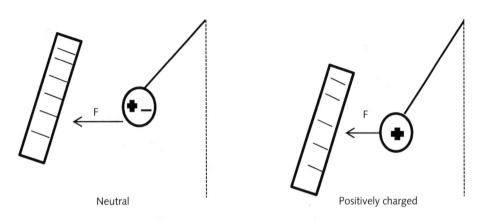

Neutral   Positively charged

Figure 14.5 Attraction of a pith ball

explains how the charged amber (when rubbed with silk) was able to attract the neutral pieces of paper. Charge polarization is only temporary.

The testing of electrostatic charges often involves the use of pith balls and electroscopes. A *pith ball* (Figure 14.5) is a small sphere (usually made or cork or *pith*) suspended on a string. If it is neutral, it will be attracted (because of charge polarization) to the charged rod; the relative strength of the force can be observed by the angle the pith ball string makes with the vertical. If the pith ball is charged, then it will either be attracted or repelled by another charged object.

An *electroscope* (Figure 14.6) can also be used to study electrostatic charges. Two types of electroscopes are generally used. One, called a leaf electroscope, uses a pair of thin metal foil sheets that will repel or attract depending on how (or if) they are charged. The other type of electroscope uses a rotating arm (or *vane*) that experiences a torque due to the presence of electrostatic charges. Both electroscopes can demonstrate the principles of charging by contact, conduction, and induction.

 Misconceptions

It is important to remember that only electrons are transferred in solids. The principle of the conservation of charge demands that the gain of electrons in one object be accompanied by an equal loss of electrons in another. You must be able to distinguish between insulators and conductors as well as the three methods of charging (contact, conduction, and induction). Grounding is an important concept, as well as the operation of electroscopes and the use of pith balls to test for the presence of net electrostatic charges. The unit of charge is the coulomb, which represents the accumulation of 6.25×10^{18} elementary charges.

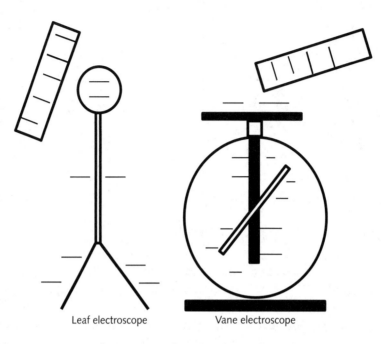

Leaf electroscope   Vane electroscope

Figure 14.6 Repulsion in an electroscope

The dynamic aspects of charges in motion (electricity or electrodynamics) will be studied in the next chapter. It is time to discuss the way in which we study the forces between charges and the amount of work done as they are separated. This is an important step in our journey of understanding electric circuits.

Coulomb's Law and Electric Fields

Charles Coulomb (in 1795) developed the relationship between two electrostatic charges at about the same time that English physicist Henry Cavendish (1731–1810) was determining the value of G (universal gravitational constant) used in Newton's Law of Gravitation:

$$\mathbf{F}_g = -G[m_1 m_2 / r^2], \text{ where } G = 6.67 \times 10^{-11} \text{ Nm}^2/\text{kg}^2$$

Coulomb, using a torsion balance similar to the one used by Cavendish, discovered that for two electrostatic charges:

$$\mathbf{F}_e = k[q_1 q_2 / r^2], \text{ where } k = 9 \times 10^9 \text{ Nm}^2/\text{C}^2$$

Notice the similarities between the two laws. Both are *inverse square law* relationships. In the case of Coulomb's Law, however, the force can either be attractive or repulsive (depending on the sign of the charges). Gravitation is always an *attractive* force. The attractive or repulsive nature of the electrostatic force leads to the idea of neutralization. Electrostatic forces can be neutralized (and shielded against), but not gravitational forces (every mass is attracted to every other mass in the universe).

Problem   Calculate the electrostatic force between a charge of $+6 \times 10^{-5}$ C and a charge of -5×10^{-6} C that are separated by a distance of 0.2 m.

Solution

Step 1. All the information is given for a direct substitution with units:

$$\mathbf{F}_e = (9 \times 10^9 \text{ Nm}^2/\text{C}^2)(+6 \times 10^{-5} \text{ C})(-5 \times 10^{-6} \text{ C})/(0.2 \text{ m})^2$$

$$= -67.5 \text{ N (attraction)}$$

We can define the *electric field* as a vector quantity equal to force per unit charge:

$$\mathbf{E} = \mathbf{F}_e/q \text{ (units are newtons per coulomb: N/C)}$$

This means that $\mathbf{F}_e = \mathbf{E}q$. The direction of the electric field is, by definition, the direction that a *positive* test charge would go if placed in the field. Recall that a field is the region surrounding an object where a force can be detected. Notice the analogy with the gravitational field strength ($\mathbf{g} = \mathbf{F}_g/m$; units are N/kg). If we imagine placing negative charges on a hollow metal sphere, the charges will move to the outside of the sphere and the electric field inside the sphere would be equal to zero!

We can map the electric field using force vectors, remembering that the direction of the electric field is the direction that a positive test charge would go (see Figure 14.7 for point charge field configurations). Try to describe the field lines between two *like* charges yourself.

Problem   Calculate the *magnitude* of the force on an electron placed in an electric field of magnitude 500 N/C.

Solution

Step 1. This problem is set up for direct substitution with units:

$$\mathbf{F}_e = \mathbf{E}q = (500 \text{ N/C})(1.6 \times 10^{-19} \text{ C}) = 8.0 \times 10^{-17} \text{ N}$$

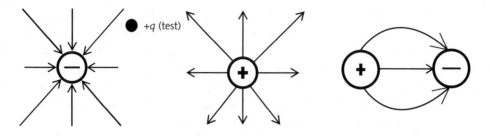

Figure 14.7 Some electric field configurations for point charges

The electric field between two charged parallel plates is uniform (constant everywhere; see Figure 14.8). Remember, the direction of the field is based on a *positive* test charge!

American physicist Robert Millikan (1868–1953) performed an experiment in 1909 using suspended, charged oil drops between parallel plates to determine the charge of an electron (and demonstrated that charges are based on whole multiples of elementary charges). By placing a negatively charged oil drop between the plates, as shown in Figure 14.8, the force of gravity ($m\mathbf{g}$) would tend to pull the drop downward (while the electric force would attract the drops upward) with a magnitude equal to $\mathbf{E}q$. If the drops were balanced, then $\mathbf{F}_g = \mathbf{F}_e$ and

$$\mathbf{E}q = m\mathbf{g}, \text{ which implies that } q = m\mathbf{g}/\mathbf{E}$$

Millikan, a meticulous experimental physicist, later confirmed the theory of the *photoelectric effect* (see Chapter 19) that was developed by Albert Einstein (1879–1955). The electron, discovered in 1897 by English physicist J. J. Thomson (1856–1940), has a known mass of 9.1×10^{-31} kg, and the proton has a known mass of 1.67×10^{-27} kg.

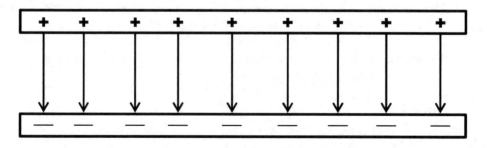

Figure 14.8 Electric field between two charged parallel plates

Misconceptions

It is important to remember that the direction of the electric field is based on the direction that an imaginary positive test charge would go. The field near point charges would increase as you get closer to it and weaker (following an inverse square law) as you move away from it. If we have two charges q_1 and q_2 separated by a distance r, the electric field acting on q_2 (due to q_1) would be given by

$$E = kq_1/r^2$$

and

$$F = Eq_2$$

The field between two parallel plates is uniform, which means the force on a charge would be the *same* everywhere. This concept will be very important as we develop the concept of *voltage* and study the development of the battery.

Potential Difference and Voltage

It takes *work* to separate unlike charges. It takes work to bring like charges together. When dealing with electrostatic charges and fields, we introduce the concept of *electric potential*. As an analogy, imagine three unequal masses sitting on a table at a distance h from the floor (see Figure 14.9). Although each mass has a different potential energy (due to the gravitational field), the potential energy per mass for each one is exactly the same! That is, while $m_1gh \neq m_2gh \neq m_3gh$, we see that $(PE_1/m_1) = (PE_2/m_2) = (PE_3/m_3) = gh$.

We call the *gravitational potential energy per mass* the *gravitational potential*. In a similar way, the *electrical potential energy per charge* is called the *electric potential*. If we compare the work done (recall that $W = \Delta PE$) to move a unit of charge from one point in the electric field to another (with different potential energies), we define a concept known as the *electric potential difference* (or *voltage*). The units would be equal to joules per coulomb (or *volts*, V). As an equation, we would write $V = W/q$ and, therefore, $W = Vq$.

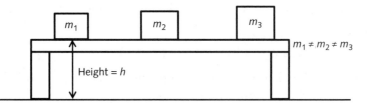

Figure 14.9 Gravitational potential energy per unit mass

Problem   What is the value of the electrical potential difference if it takes 10 J to move 2 C of charge through an electric field?

Solution

Step 1. The problem is set up for direct substitution with units:

$$V = (10 \text{ J})/(2 \text{ C}) = 5 \text{ V}$$

For the movement of charge between two parallel plates (where the electric field is uniform), the work done per charge is expressed in terms of a *potential gradient*. Suppose we have two charged parallel plates separated by a distance d. Then, since the force on a charge moved through the uniform field is constant (see Figure 14.8),

$$\mathbf{F} = \mathbf{E}q$$

which means

$$\mathbf{F}d = \mathbf{E}qd$$

and, therefore,

$$(\mathbf{F}d)/q = \text{work done per charge} = \text{voltage } (V) = \mathbf{E}d$$

$$V = \mathbf{E}d \quad \text{and} \quad \mathbf{E} = V/d \text{ (in units of volts per meter)}$$

The quantity V/d is called the *potential gradient*. You should note that the units of V/m are equivalent to N/C!

Problem   Two parallel plates are connected to a 100-V source of potential difference. If they are separated by a distance of 0.02 m, then calculate the strength of the electric field between the plates and the magnitude of the force exerted on an electron that is moving through the field.

Solution

Step 1. To find the strength of the electric field between the parallel plates, we use:

$$\mathbf{E} = V/d = (100 \text{ V})/(0.02 \text{ m}) = 5000 \text{ V/m} = 5000 \text{ N/C}$$

Step 2. The magnitude of the force on an electron moving through the field (which is uniform) is given by

$$\mathbf{F} = \mathbf{E}q = (5000 \text{ N/C})(1.6 \times 10^{-19} \text{ C}) = 8.0 \times 10^{-16} \text{ N}$$

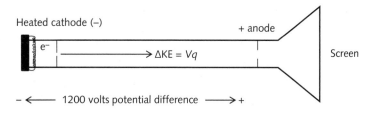

Figure 14.10 Thermionic emission

Electrons can be emitted from a heated filament (called a cathode) and then accelerated from rest to a maximum velocity in a uniform electric field (see Figure 14.10). This process is known as *thermionic emission*. Simple devices that often use thermionic emission in this way are called *cathode ray tubes*. The kinetic energy gained by the electrons (e⁻) is equal to the work done by the potential difference between the cathode (negative plate) and the anode (positive plate):

$$\Delta KE = \frac{1}{2}m\mathbf{v}^2 = Vq \text{ (units in joules)}$$

We define a new unit of energy called the *electron-volt* (eV) as follows: *1 electron-volt is equal to the energy given to one electron (or elementary charge) accelerated from rest by a potential difference of 1 volt.* Hence, if one electron is accelerated in a cathode ray tube by a potential difference of 1200 V, it gains 1200 eV of kinetic energy. Using the known charge on an electron, it is easy to see that 1 eV = 1.6 × 10⁻¹⁹ J. Thus, 1200 eV = 1.92 × 10⁻¹⁶ J. Using the known mass of the electron (9.1 × 10⁻³¹ kg), we can determine the velocity of electrons in these devices. In our example:

$$1.92 \times 10^{-16}\,J = \frac{1}{2}(9.1 \times 10^{-31}\,kg)\mathbf{v}^2$$

and

$$\mathbf{v} = 2.05 \times 10^7 \text{ m/s}$$

Misconceptions

It is important to remember that the potential difference (or voltage) is a measure of the work done *per charge* ($\Delta PE/q$) in units of J/C or volts. This is the essence of the battery, which stores potential difference using two different metals (such as lead and zinc) and an acid to produce an electrochemical reaction. We will not go into the details of the chemistry involved in battery design here.

Finally, we do not have space to go into all details of electrostatics, but you should know that a simple device for storing static charges (using the design of parallel plates and an insulating material known as a *dielectric*) is called a *capacitor*. Capacitors play a very important role in advanced circuits, but we will be discussing only simple electric circuits in the next chapter. The capacitor is quantified by its *capacitance*, *C* (its ability to store charge when connected to a voltage source), in units of *farads* (**F**) such that

$$C = q/V$$

More information about capacitors can be found in more advanced physics books.

Problem-Solving Strategies to Avoid Missteps

As you solve problems with electrostatic forces and fields, it is always important to remember (as we have repeated many times) that Coulomb's Law is an inverse square law. This means you must make sure not only to use the correct units but to remember the force varies inversely with the square of the distance between the two charges. Additionally, even if you have electrons moving in free space, the direction of the electric field is the direction that a positive test charge would go. Finally, the voltage source is referred to as a source of potential difference equal to the work done per unit coulomb of charge (1 C = 6.25×10^{18} elementary charges). The problem-solving ring is always available for you to consult as you plan a problem-solving strategy.

Exercise 14.1

1. Three charges are arranged along a line. On the left, a charge (A) of 7.0×10^{-6} C is fixed. A second charge (B) of -2.0×10^{-6} C is located 0.5 m to the right of charge A. Finally, a third fixed charge (C) of 3.0×10^{-6} C is located 0.3 m from charge B. Calculate the *net* force acting on charge B due to the two fixed charges.

2. A proton is accelerated from rest by a potential difference of 800 V between two parallel plates. Calculate the kinetic energy gained by the proton in units of eV and joules. Using the known mass of a proton (1.67×10^{-27} kg), calculate the velocity gained by the proton.

3. Two parallel plates are separated by 0.03 m and connected to a 90-V potential difference. Calculate the strength of the electric field between the two plates. Calculate the force exerted on an electron placed in that field.

4. Two metal spheres are placed on insulating stands. One sphere has a charge of 6 C while the second sphere has a charge of −4 C. If the two spheres are touched, calculate the new charge on *each* sphere.

15

Electric Circuits

In this chapter you will learn about the operation of simple electric circuits. The key concepts are *electric current* and *Ohm's Law*. Electricity powers the modern world. Understanding the structure of simple circuits is the next step in our journey.

Ohm's Law

A simple circuit consists of a source of potential difference and a closed conducting loop. A battery separates charges, so when the ends of the battery are connected to a wire, charge will flow. If we consider the flow of electrons through a conductor, then the direction of the current will be from the negative terminal around toward the positive terminal.

We define the electric *current* to be equal to the rate of flow of charge:

$$I = \Delta q/\Delta t \text{ (units are coulombs per second or amperes, A)}$$

For example, if we have 50 C of charge passing a given point in a circuit every 5 s, then the current would be equal to 10 A. A device that measures current is called an *ammeter*. A device that measures potential difference is called a *voltmeter*.

In some cases, we imagine the flow of positive charge in a circuit, in which case we talk about the *conventional current*. Some textbooks define current to be the net positive flow in a circuit, while others define current as the flow of electrons through the circuit. To be consistent, we will use the word *current* to mean the conventional current of positive charge flow.

To better understand simple circuits, we will need to introduce some schematic symbols that are used to draw them (see Table 15.1).

A simple circuit consisting of a light bulb, a battery, a switch, a voltmeter, and an ammeter would look like Figure 15.1.

Table 15.1 Schematic Symbols for Electric Circuits

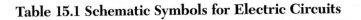

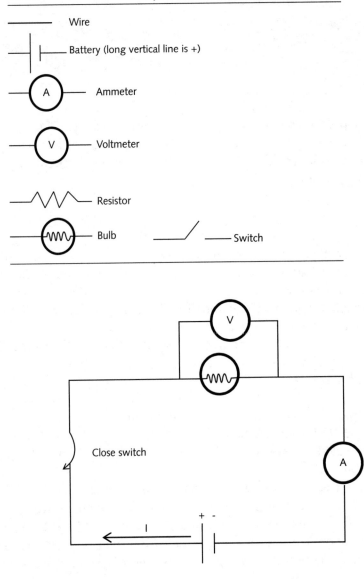

Figure 15.1 A simple circuit

In our simple circuit, we have shown the conventional (+) current that would flow once the switch is closed. Notice how the meters are placed. The ammeter is placed within the circuit (*in series*) since it measures the flow of charge. The voltmeter is connected across (*in parallel*) the bulb (load or resistor) since it is measuring the *potential drop* across that part of the circuit.

The electrical power generated (at constant temperature) by the circuit is given by:

$$P = VI \text{ (units are in watts)}$$

The power is measured in units of joules per second (watts), so the energy produced by doing work on charges during a time interval t (in seconds) would be given by:

$$W = Pt = VIt \text{ (units are in joules)}$$

Problem   A simple circuit is connected with a 12-V battery and records a current of 2 A. How much power is generated? If the circuit is on for 10 s, how much energy is used?

Solution

Step 1. To find the power, we simply multiply the voltage times the current:

$$P = VI = (12 \text{ J/C})(2 \text{ C/s}) = 24 \text{ J/s} = 24 \text{ W}$$

Step 2. The energy generated in 10 s is equal to the work done:

$$W = Pt = (24 \text{ W})(10 \text{ s}) = (24 \text{ J/s})(10 \text{ s}) = 240 \text{ J}$$

It was discovered that if the voltage is changed (at constant temperature), then there will be a directly proportional change in the current. This was first developed mathematically by German physicist Georg Simon Ohm (1789–1854) and studied experimentally by English physicist Charles Wheatstone (1802–1875). The relationship is known as *Ohm's Law* and is written in the form:

$$R = V/I \text{ (units are Ohms, } \Omega)$$

The quantity R (which would be the slope of a line in a graph of voltage versus current) represents the *electrical resistance* of the circuit. Ohm's Law is also written in the form:

$$V = IR$$

This means that the power can be written as:

$$P = VI = I^2R = V^2/R$$

and

$$W = VIt = I^2Rt = (V^2/R)t$$

Problem   For the circuit in the sample problem (12 V of potential difference and 2 A of current), calculate the resistance of the circuit (assume constant temperature).

Solution

Step 1. We simply use Ohm's Law:

$$R = V/I = (12 \text{ V})/(2 \text{ A}) = 6 \text{ }\Omega$$

If the temperature of the resistor increases, then its resistance will increase due to the increase in the molecular motion of the conductor. However, at constant temperature, there are several factors that influence the resistance of a conductor. In general, if the length of the conductor is increased, then the resistance will proportionally increase (because there is more material for the electrons to move through). If the thickness (cross-sectional area) changes, then the resistance will change inversely because a thicker wire will have more room for the electrons to flow. Finally, different materials are characterized by their resistivity (in units of $\Omega \cdot m$) at constant temperature (typically 20°C). These results can be experimentally verified and written in the form:

$$R = \rho L/A \text{ (where } \rho = \text{ resistivity at 20°C)}$$

A sample of some typical resistivities is given in Table 15.2.

Table 15.2   Resistivities at 20°C

MATERIAL	RESISTIVITY ($\Omega \cdot M$)
Aluminum	2.82×10^{-8}
Copper	1.72×10^{-8}
Gold	2.44×10^{-8}
Nichrome	150.0×10^{-8}
Silver	1.59×10^{-8}
Tungsten	5.60×10^{-8}

Problem Calculate the resistance (at 20°C) for 3 m of copper wire that has a diameter of 0.20 mm.

Solution

Step 1. To solve this problem we first need to find the cross-sectional area. We assume that a wire has a circular cross-sectional area of $\pi r^2 = \pi D^2/4$ (where D = diameter in meters).

$D = 0.20$ mm $= 2 \times 10^{-4}$ m

$A = \pi(2 \times 10^{-4}$ m$)^2/4 = 3.14 \times 10^{-8}$ m^2

Step 2. From the chart, we know the resistivity of copper is $\rho = 1.72 \times 10^{-8}$ $\Omega \cdot$ m, therefore:

$R = (1.72 \times 10^{-8}\ \Omega \cdot$ m$)(3$ m$)/(3.14 \times 10^{-8}$ m$^2) = 1.64\ \Omega$

Series Circuits

A series circuit consists of several resistors connected in sequence, one after the other. This has the effect of increasing the length of the conductor, so we know that the total resistance increases as the number of resistors in series are increased. A simple series circuit is shown in Figure 15.2.

Experiments show that the current in the series circuit made up of one loop is the same everywhere. However, because there are potential drops across each resistor, the total voltage across the battery is equal to the sum of

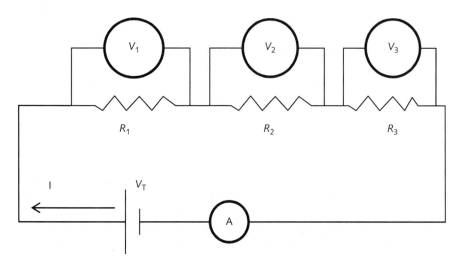

Figure 15.2 A simple series circuit

the potential drops across each resistor. Additionally, if one resistor (let's say it is a light bulb) *goes out*, then the whole circuit *goes out*.

Using Ohm's Law and our experimental results, we find that:

$I = I_1 = I_2 = I_3 = \ldots$ V_T = Total voltage (potential difference)

$V_T = V_1 + V_2 + V_3 + \cdots$ $R_{eq} = V_T/I$ (equivalent resistance)

$R_T = R_{eq} = R_1 + R_2 + R_3 + \cdots$

Problem Resistors of 10 Ω, 15 Ω, and 5 Ω are all connected in series to a 90-V source of potential difference. Calculate the total equivalent resistance of the circuit, the total current in the circuit, and the potential drop across each resistor.

Solution

Step 1. The goal of the problem is to analyze a series circuit. To find the equivalent resistance, all we need to recognize is that the total resistance is equal to the sum of the individual resistors (in series):

$R_{eq} = 10\ \Omega + 15\ \Omega + 5\ \Omega = 30\ \Omega$

Step 2. Using Ohm's Law, we can easily find the total current in the circuit:

$I = V_T/R_{eq} = 90\ V/30\ \Omega = 3\ A$

Step 3. To find the potential drop across each resistor, we again use Ohm's Law:

$V_1 = I_1R_1$ $V_2 = I_2R_2$ $V_3 = I_3R_3$ (but $I = I_1 = I_2 = I_3 = \ldots$)

$V_1 = (3\ A)(10\ \Omega) = 30\ V$ $V_2 = (3\ A)(15\ \Omega) = 45\ V$ $V_3 = (3\ A)(5\ \Omega) = 15\ V$

Notice that all three potential drops add up to 90 V.

Misconceptions

In reality, there is some energy loss and a voltmeter might record potential drops that are lower. In this way, we often speak of EMF (which stands for *electromotive force*, but it is *not* a force) instead of potential difference to allow for the transfer of kinetic energy per charge as the charges actually flow through the circuit. Also, the analogy of electron flow to water flow is not accurate, as the motion of electrons through a conducting wire does not make a smooth path like water through a pipe. The water flow analogy is useful to some extent, but like most analogies, you must be careful when applying it.

Parallel Circuits

Oftentimes, a circuit will have a branch or junction. The analysis of complex circuits in the nineteenth century led to many modern innovations. The German physicist Gustav Kirchoff (1824–1887) developed the junction rule, which states that the sum of the currents entering a branch (or junction) must be equal to the sum of the currents leaving a branch (or junction). This is illustrated in Figure 15.3 (the black dot represents the branch point). The concept of a circuit branch is essential to understanding parallel circuits.

A simple parallel circuit would look like Figure 15.4. The circuit current I would split into two currents, I_1 and I_2 (as measured by ammeters A$_1$ and A$_2$).

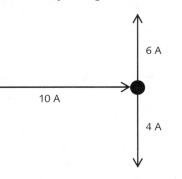

Figure 15.3 Kirchoff's junction rule (notice the direction of currents)

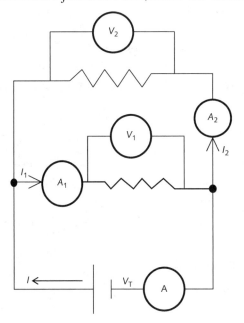

Figure 15.4 A simple parallel circuit

Those two currents must add up to the total circuit current I coming from the battery. However, experiments show that the potential drops across each resistor in parallel are the same.

Thus, we have for a parallel circuit:

$$V_T = V_1 = V_2 = V_3 = \dots$$
$$I = I_1 + I_2 + I_3 + \dots$$

Now, using Ohm's Law:

$$\frac{1}{R_{eq}} = \frac{1}{R_1} + \frac{1}{R_2} + \frac{1}{R_3} \quad \text{(the resistors add reciprocally!)}$$

Misconceptions

The effect of adding a branch is to reduce the equivalent resistance of the circuit. This increases the overall circuit current. Additionally, the extra branches mean that if one part of the circuit goes out, then the rest of the circuit can stay on (with a closed conducting path). It is very important for you to know the differences between series and parallel circuits. In a parallel circuit, the equivalent resistance will always be *lower* than the lowest resistor present. Don't forget to take the reciprocal at the end since the equation is for $1/R_{eq}$! You should also be familiar with how to draw them schematically.

Problem Two resistors with resistances of 5 Ω and 20 Ω each are connected in parallel to a 16-V battery. Calculate the equivalent resistance of the circuit, the total circuit current, and the current in each branch of the circuit.

Solution

Step 1. This is a parallel circuit; therefore, to find the equivalent resistance, we need to follow the equation and add reciprocal resistances (using the least common denominator):

$1/R_{eq}$ = 1/5 Ω + 1/20 Ω = 5/20 Ω (don't forget to take reciprocal to find R_{eq}!)

which means that R_{eq} = 4 Ω.

Step 2. Using Ohm's Law, we know that $I = V_T/R_{eq}$ = (16 V)/(4 Ω) = 4 A.

Step 3. In a parallel circuit, the voltage drop is the same across each resistor. To find each branch current, we simply use Ohm's Law on each branch:

$I_1 = (16 \text{ V})/(5 \text{ } \Omega) = 3.2 \text{ A}$ and $I_2 = (16 \text{ V})/(20 \text{ } \Omega) = 0.8 \text{ A}$

Notice that both currents add up to 4 A (as they should) and that the current in the 5-Ω resistor is four times larger than the current in the 20-Ω resistor (as expected).

Problem-Solving Strategies to Avoid Missteps

When solving problems with electric circuits it is important to make sure you understand how a circuit diagram is schematically drawn, including the placement of the ammeter and voltmeter. Make sure you understand the differences between series and parallel circuits. In a series circuit, the current is the same everywhere and the voltage drops are shared. In a parallel circuit, the voltage drops are the same everywhere, but the currents are split according to Kirchoff's junction rule. Ohm's Law is one of the main concepts as are the factors affecting the resistance of a conductor. Finally, make sure to identify whether the conventional current or electron flow current is being used.

Exercise 15.1

1. Calculate the resistivity of 30 m of conducting wire that has a diameter of 0.0035 m and a measured resistance of 0.50 Ω. Assume a standard temperature of 20°C.

2. How much energy is generated by a 1500-Ω resistor connected to a 110-V source of potential difference for 5 minutes?

3. Three resistors of 50 Ω, 30 Ω, and 40 Ω are connected in series to a 60-V battery. Calculate the equivalent resistance of the circuit, the total current in the circuit, the total power generated by the circuit, and the voltage drop across each resistor.

4. Three resistors of 30 Ω, 15 Ω, and 10 Ω are connected in parallel to a 20-V battery. Calculate the equivalent resistance of the circuit, the total current for the circuit, and the current flowing through each resistor.

16

Magnetism

In this chapter you will learn about magnetism and its relationship to electricity. The unification of electricity and magnetism (electromagnetism) is a very important step in our continuing journey of exploration. Electric motors, transformers, and most of our modern technology operate on the principles of electromagnetism. Additionally, the vibration of electromagnetic fields produces electromagnetic waves. These waves are the basis for the study of light and optics.

Terrestrial Magnetism

In the year 1600, Sir William Gilbert (1544–1603) published *De Magnete* and revolutionized the study of terrestrial magnetism. Ancient people used rocks, often called *lodestones*, found in the ground to attract small bits of metal. Iron was easily attracted to these rocks, but copper was not. If an iron rod was rubbed by these rocks, it too developed these attractive properties. Iron rods that have been placed in the earth for many years also become magnetic.

Gilbert demonstrated that magnets (the name may derive from an ancient Greek region called *Magnesia* where some of these rocks were found) attract as well as repel and have two poles. He demonstrated that the earth itself was a magnet and the operation of a compass was due to the interaction between the earth's magnetic field and a magnetized iron needle. Magnetic poles are designated as north and south (with the north geographic pole of the earth actually a south magnetic pole!).

We will discuss terrestrial magnetism in terms of iron (or ironlike compounds). On the periodic table, the elements iron, cobalt, and nickel exhibit the greatest magnetic properties. These materials are known as *ferromagnetic* substances (we will not be studying other properties known as paramagnetism or diamagnetism). We use the word *permeability* to classify the magnetic properties of materials (we will also not be going into detail about this).

Ferromagnets, such as a typical bar magnet, can be studied using sprinklings of iron filings or a compass. The force of attraction (or repulsion) can be used to map out the magnetic field. However, unlike electrostatic fields (which can originate from point elementary charges), magnets are *dipoles* (they always have a north and a south pole). We do not have an isolated *test pole* to map the field (recall that the electrostatic field was mapped using an imaginary positive test charge), so a compass is used instead (see Figure 16.1).

The field surrounding a bar magnet goes from north to south and the compass needle aligns itself tangent to the field (a torque is produced when the compass is first placed in the field). Charles Coulomb investigated the force law between magnets, but we will use the concepts of electromagnetism instead to define the force law and measure field strength.

Various other magnetic fields can be seen in Figure 16.2 (again, fields go from north to south).

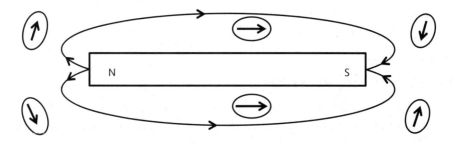

Figure 16.1 The electromagnetic field surrounding a bar magnet

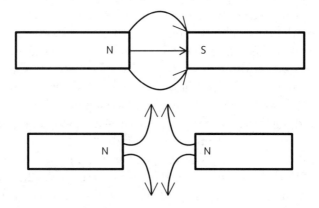

Figure 16.2 Other magnetic fields for opposite and like poles

The relative strength of the field near a bar magnet can be observed from the concentration density of the lines of force (magnetic flux density) as seen using iron filings. You can also use a compass and measure the deflecting effect as one gets closer to the magnet. It is more common, however, to use the concepts of electromagnetism, to which we now proceed.

Electromagnetism

In 1819, Danish physicist Hans Christian Oersted (1771–1851) demonstrated that an electric current can deflect a compass. The shape of the electromagnetic field was a three-dimensional circle around a wire (see Figure 16.3, which demonstrates conventional current and the so-called *right-hand rule*). Because a compass needle deflects due to the presence of an external magnetic field (in this case, electromagnetic), the amount of deflection can be related to the strength of the field. Conceptually, it can be demonstrated that the strength of the field depends directly on the amount of current in a long, straight wire, on how many wires there are, and inversely on the distance from the wire. The symbol **B** is used to represent the magnetic field strength (it is a vector quantity). The positioning of the compass shows the field (using positive conventional current) to be circular and its direction (using a right-hand rule). If *electron flow* current is used, then the direction would follow the *left-hand rule.*

Right-Hand Rule 1

For a straight wire with *conventional* current flowing, the thumb of the right hand points in the direction of the current and the fingers curl in the

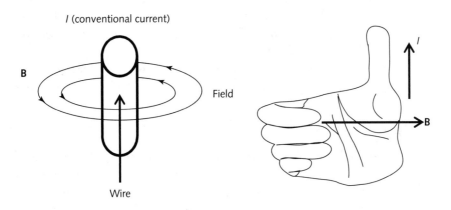

Figure 16.3 Right-hand rule 1 for a long, straight wire with conventional current

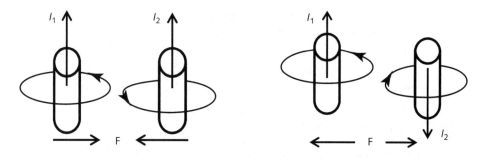

Figure 16.4 Forces on current-carrying wires (perspective sketches)

direction of the electromagnetic field, **B**. If two wires with conventional current flowing through them are placed near each other, then there will be a force produced due to the interaction of their electromagnetic fields. As shown in Figure 16.4, if the currents are going in the *same* direction, then an *attractive* force will be produced. If the currents are flowing in *opposite* directions, then a *repelling* force will be produced.

The fields are three-dimensional and schematic symbols help in understanding the perspective drawings. Let

X = a vector field going *into* the page
• = a vector field coming *out* of the page

If wire is looped around an iron object (or another highly permeable core material), then a shape called a *solenoid* (see Figure 16.5) is produced. When connected to a battery, the solenoid becomes an electromagnet and the magnetic field looks just like the field produced by a bar magnet. The strength of the electromagnetic field will depend directly on the permeability of the core, the number of turns of wire, and the conventional current in the wire and will be inversely proportional to the radii of the loops. We could also have a circular loop of wire.

The direction of the electromagnetic field (**B**) is determined from a second right-hand rule (or left-hand rule if electron flow current is used).

Right-Hand Rule 2

For a solenoid (or coil of wire) with conventional current flowing through it, the fingers of the right hand curl in the direction of the current, and the thumb points toward the north pole of the electromagnetic field.

When a wire with current is placed in an external magnetic field, a force is induced on the wire. Whether the direction of the current is perpendicular

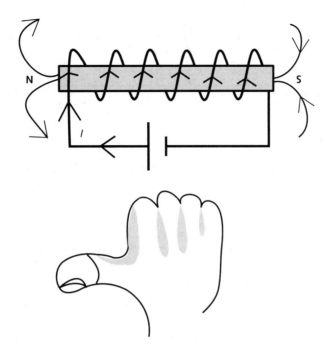

Figure 16.5 Right-hand rule for a solenoid (conventional current)

to the direction of the magnetic field depends on the strength of the field (**B**), the current in the wire, and the length of wire in the field. From this, we can demonstrate that

$$\mathbf{F} = \mathbf{B}IL$$

Since the force is measured in newtons, this equation can be used to define the strength of the electromagnetic field:

$$\mathbf{B} = \mathbf{F}/IL$$

where the units are N/A · m (also called teslas, T).

The direction of the force is given by a third right-hand rule (assuming that the direction of the conventional current is perpendicular to the external magnetic field (see Figure 16.6 for a magnetic field directed *into* the page).

Right-Hand Rule 3

For a long, straight wire with conventional current flowing through it, perpendicular to an external magnetic field, the direction of the induced force

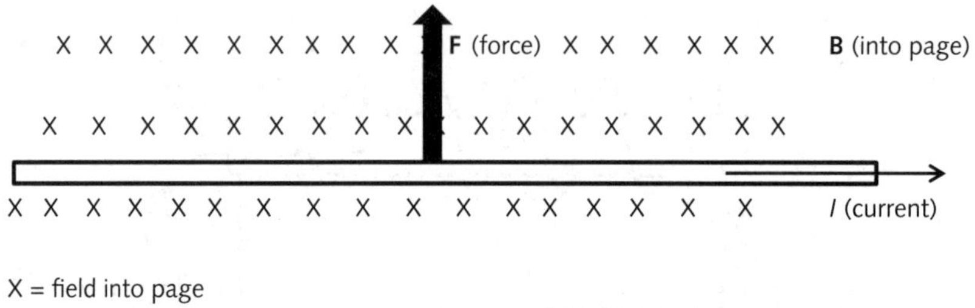

X X X X X X X X X X **↑ F** (force) X X X X X X   **B** (into page)

X X X X X X X X X X X X X X X X X

X X X X X X X X X X X X X X X X X X   **I** (current)

X = field into page

Figure 16.6 Force induced on a straight wire with conventional current in an external field

on the wire is mutually perpendicular to both the current and the field. If the fingers of the right hand point along the direction of the field and the thumb of the right hand points along the direction of the conventional current, then the direction of the open palm indicates the direction of the induced force on the wire.

If we have a beam of charged particles crossing a magnetic field at constant velocity, then since $I = q/t$,

$$\mathbf{F} = \mathbf{B}IL = \mathbf{B}(q/t)L = \mathbf{B}q\mathbf{v}$$

The direction of this force would be determined by right-hand rule 3 (for positive charges) or a corresponding left-hand rule for negative charges.

Problem Calculate the magnitude of the force on a beam of electrons crossing perpendicularly to a magnetic field of 0.03 T with a velocity of 3.5×10^6 m/s.

Solution

Step 1. For this problem, all we have to do is substitute in the appropriate values:

$$\mathbf{F} = \mathbf{B}q\mathbf{v} = (0.03 \text{ T})(1.6 \times 10^{-19} \text{ C})(3.5 \times 10^6 \text{ m/s}) = 1.68 \times 10^{-14} \text{ N}$$

Misconceptions

Hand rules can be confusing. It is very important to memorize all of them. Also, knowing when to use the right hand (conventional current) versus the left hand (electron flow current) can be tricky. Read all of the information in a problem carefully. Most textbooks now use *conventional* current!

Problem A 2.5-m wire has 5 A of current flowing through it. When it is placed perpendicular to a magnetic field, a force of 1.5 N is exerted on it. Calculate the strength of the magnetic field.

Solution

Step 1. For this problem, we use the equation that defines the magnetic field strength:

$$\mathbf{B} = \mathbf{F}/IL = (1.5 \text{ N})/(5 \text{ A})(2.5 \text{ m}) = 0.12 \text{ T}$$

If the velocity of a moving charge (with mass m and charge q) is perpendicular to an external magnetic field, then it is possible to have the induced magnetic force deflect the charge into a circular path. Under these conditions, the magnetic force (which is a net force) acts at right angles to the velocity and so our old friend centripetal force is at play. Thus,

$$m\mathbf{v}^2/r = \mathbf{B}q\mathbf{v}$$

and

$$m\mathbf{v} = \mathbf{B}qr$$

If the mass and velocity are unknown, then a second experimental method can be used. We know from our study of cathode ray tubes that the kinetic energy given to an electron (or any charge) is proportional to the accelerating voltage V:

$$\mathbf{KE} = Vq \text{ (where } q \text{ is the elementary charge on an electron)}$$

We also know that kinetic energy is given by:

$$\mathbf{KE} = \frac{1}{2}m\mathbf{v}^2$$

and, therefore,

$$Vq = \frac{1}{2}m\mathbf{v}^2$$

We now have two equations and two unknowns to solve for the mass and velocity of our charges independently. If we eliminate the velocity for a charge accelerated by a voltage V, then we have a device known as a *mass spectrograph*:

$$m = \mathbf{B}^2 q r^2 / 2\ V$$

Problem   A doubly ionized ion (loss of two elementary charges) is moving perpendicularly to a magnetic field of 0.05 T in a mass spectrograph after being accelerated by a potential difference of 300 V. If the observed radius of the path is 0.85 m, calculate the mass and velocity of the ion.

Solution

Step 1. The goal of this problem is to find the mass and velocity of the ion. To find the mass, we use the equation we derived for the mass spectrograph. Note that the charge is twice the elementary charge value (doubly ionized):

$$m = (0.05\ \text{T})^2 (3.2 \times 10^{-19}\ \text{C})(0.85\ \text{m})^2 / 2(300\ \text{V}) = 9.63 \times 10^{-25}\ \text{kg}$$

Step 2. To find the velocity, we can now use the equation:

$$m\mathbf{v} = \mathbf{B}qr$$

$$\mathbf{v} = (0.05\ \text{T})(3.2 \times 10^{-19}\ \text{C})(0.85\ \text{m})/(9.63 \times 10^{-25}\ \text{kg}) = 1.4 \times 10^5\ \text{m/s}$$

The concepts of forces induced on moving charges in magnetic fields have many practical applications. These range from medical imagining, to computer monitors, to electric motors, and finally to the study of the fundamental properties of matter.

Electromagnetic Induction

After the discovery of electromagnetism by Oersted in 1819, the quest for the reverse process began. If moving electric charges can produce a magnetic field, can a changing magnetic field produce an electric current? Michael Faraday found the answer around 1831. By moving a wire connected to a microammeter (also known as a galvanometer) across a magnetic field, Faraday was able to produce a small electric current.

The effect (now called *electromagnetic induction*) can be enhanced if a loop (or coil) of wire is used. By moving a magnet into the coil, it was discovered that current flows in one direction. By moving the magnet out of the coil, the current reverses direction. The same effect occurs if the coil moves over the magnet (relative motion is the key).

Current flows when a potential difference exists. However, with induction there is no stored potential difference. Instead, it is EMF (electromotive force, measured in volts) that drives the current. You can think of the EMF as the change in kinetic energy per unit charge (in the same way that potential difference was the change in potential energy per unit charge). How does this happen? Looking at Figure 16.7, we can use right-hand rule 3 to illustrate what happens. In this case, you would let your thumb (of your right hand) represent the direction of the velocity (remember, for electron flow current you would use your *left* hand!).

As the wire is moved perpendicularly through the field (cutting the field lines), the electrons that exist in the wire are moved through the field also. Assume that a length of wire L is being moved as shown with velocity **v**. Then if we write:

$$\mathbf{F} = \mathbf{B}q\mathbf{v}$$

the work done (ΔKE) by the force on the electrons through the length L is given by

$$\Delta KE = \mathbf{F}L = \mathbf{B}qvL$$

Therefore,

$$EMF = \mathbf{B}vL$$

The induced EMF (which is not stored) is proportional to the strength of the external field, the length of the wire in the field, and the velocity of the wire. This equation expresses Faraday's Law of Induction: *The induced EMF in a conductor is directly proportional to the rate of change of magnetic flux.*

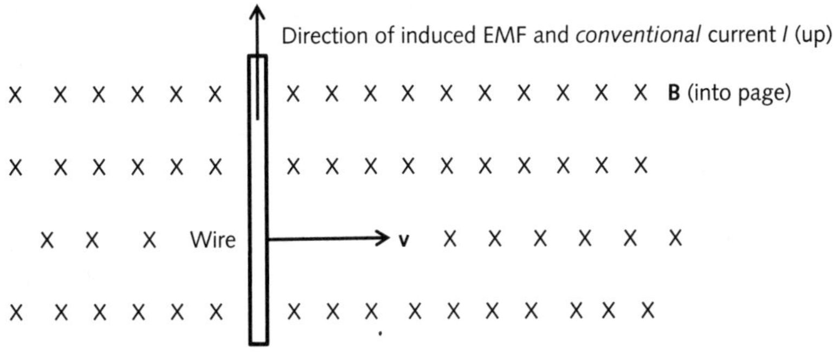

Figure 16.7 Induction in a wire moved across a magnetic field directed into the page

What is magnetic flux? Let's quickly (and briefly) examine the units of the EMF (which are expressed in volts) to assist us. Using dimensional analysis, we see that for the induced EMF the units can be written in the form:

$$\mathbf{B}vL \rightarrow (T)(m/s)(m) = (T \cdot m^2)/s$$

The quantity expressed in the units $T \cdot m^2$ corresponds to the product of the magnetic field ($\mathbf{B}$) in an enclosed area (A). This quantity, $\mathbf{B} \cdot A$, is called the magnetic flux ($\mathbf{\Phi}$). Faraday's Law is sometimes expressed in the form:

$$EMF = -\Delta\mathbf{\Phi}/\Delta t$$

(The negative sign is needed because of Lenz's Law, discussed later.) We will not be discussing the concept of magnetic flux in any more detail.

Additionally, the magnetic field can be thought of as *magnetic flux density* (based on the concentration of field lines). The unit for magnetic flux is called the *weber* (Wb). Magnetic field strength could also be expressed in units of webers per square meter (Wb/m²). We will continue to use the more convenient (and common) unit of teslas for magnetic field strength.

Problem   Calculate the induced EMF in a wire 1.2 m long moving with a velocity of 20 m/s perpendicularly through a magnetic field of 0.08 T.

Solution

Step 1. The EMF can be found by a simple substitution:

$$EMF = \mathbf{B}vL = (0.08\ T)(20\ m/s)(1.2\ m) = 1.92\ V$$

From Faraday's discovery, we can see that the only way to maintain a continuous flow of charge is to have an alternating change in magnetic flux. This generates an alternating current (AC) in the conductor. However, looking at Figure 16.7 we see something interesting. If the conventional current is induced up the wire (as shown), then once the induced current begins to flow that current itself induces a magnetic field. In fact, if you use the third right-hand rule to find the induced current (going up the wire), you will notice a force produced to the *right* that opposes the motion of the wire (which was toward the *left*).

This opposition was investigated by Russian physicist Heinrich Lenz (1804–1865) and is known as Lenz's Law. In one version, the law states that *an induced current in a conductor always flows in a direction such that its magnetic field opposes the magnetic field that induced it.* The negative sign commonly seen in the mathematical form of Faraday's Law is due to this opposition to the changing magnetic flux. This is essentially just a restatement of the Law of the Conservation of Energy, which prevents an infinite buildup of energy by producing a back-EMF in the conductor.

Faraday's Law does not require motion to be effective. As long as we have a changing magnetic field (or magnetic flux), induction will take place. The use of alternating current assists in this fact.

An application of this concept can be observed in electromagnetic transformers. Imagine a solenoid connected to an AC generator ⎯Ⓝ⎯. It has a number of turns of wire, N_p, and the effective AC voltage is given by V_p (the letter p stands for *primary*). This system can cause induction to take place in a second solenoid (the secondary) placed a distance away. If the secondary coil (with turns N_s) is connected to an AC voltmeter, then an induced secondary voltage (V_s) will be recorded (see Figure 16.8) and related to the primary voltage as follows:

$$\frac{V_s}{V_p} = \frac{N_s}{N_p}$$

If there are no losses due to heat (transformer is 100% efficient), then the power input is equal to the power output:

$$V_p I_p = V_s I_s$$

If $N_p < N_s$, then we have a *step-up* transformer. If $N_p > N_s$, then we have a *step-down* transformer.

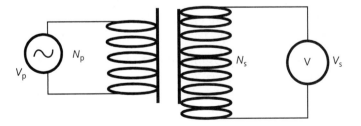

Figure 16.8 Schematic representation of a step-up transformer (AC)

Problem   An ideal step-up transformer has 50 turns on the primary coil and 800 turns on the secondary coil. If the primary AC voltage is 5 V, calculate the secondary voltage. If the secondary current is 0.5 A, what was the primary current?

Solution

Step 1. To find the secondary AC voltage, we use the information given in our equation for an ideal transformer:

$$V_s = (N_s/N_p)V_p = (800/50)(5 \text{ V}) = 80 \text{ V}$$

Step 2. Since the transformer is ideal, the power in the primary must be equal to the power in the secondary:

$$V_pI_p = V_sI_s$$
$$(5 \text{ V})I_p = (80 \text{ V})(0.5 \text{ A})$$

$I_p = 8$ A (if we step up the voltage, we must step down the current!)

Problem-Solving Strategies to Avoid Missteps

The most important concepts to master in the study of electromagnetism are the hand rules. For positive charge flow, you use the right hand, and for negative charge flow, you use the left hand. It is important to master these rules and how they are applied. Magnetic fields always are directed from north to south (there are no isolated magnetic monopoles in current elementary magnetic theory). Refer to the problem-solving ring and make sure you know the proper units. Remember, electromagnetism is due to *moving* charges. Even oscillating electromagnetic fields transfer energy, but in the form of waves (such as light).

Exercise 16.1

1. An electron ($m = 9.1 \times 10^{-31}$ kg) is moving with a velocity of 4.5×10^7 m/s perpendicularly to a magnetic field of strength 0.08 T. Calculate the radius of the circular path taken by the electron in the field.

2. A *doubly ionized* ion (loss of two electrons) is accelerated in a mass spectrometer by a potential difference of 700 V. It is deflected by a magnetic field of strength 0.06 T with an observed radius of deflection equal to 0.75 m. Calculate the mass of the ion.

3. A step-down transformer has 20,000 turns on its primary coil and 500 turns on its secondary coil. If the voltage across the secondary is 110 V, calculate the primary voltage.

4. An ion passes without deflection through a velocity selector in which a 3000 N/C electric field is perpendicular to a magnetic field of strength 0.002 T. Calculate the velocity of the ion.

5. Calculate the induced EMF in a wire of length 0.35 m placed perpendicularly to a 4.0 T magnetic field moving with a speed of 15 m/s. If the resistance of the wire is 5 Ω, calculate the induced current.

17

Properties of Light

In this chapter you will learn about the main properties of light as an electromagnetic wave. These properties include reflection, refraction, diffraction, and interference. Our study of light progresses nicely from our study of electromagnetism. It forms the basis of the study of optics (mirrors and lenses) and many other devices, such as microscopes and telescopes. It is the next step on our journey of exploration.

Electromagnetic Waves

Light is a *transverse electromagnetic wave.* James Clerk Maxwell (1831–1879) was the first to develop a mathematical model in which the oscillations of electromagnetic fields travel through space as waves with the same velocity as the velocity of light. The velocity of light (in air and a vacuum) is designated by the letter **c** and is approximately equal to 3×10^8 m/s.

The experimental verification of electromagnetic waves was developed by Heinrich Hertz (1857–1894) and many others. The importance of light being an electromagnetic wave means that it *does not* require a material medium for propagation (light is *not* a mechanical wave).

The wave nature of light had been experimentally established in 1801 by Thomas Young (1773–1829), who demonstrated that when light is passed between two narrow slits it displays a characteristic diffraction and interference pattern. The transverse nature of light waves is established through the property of *polarization.* Recall that the vibrations of a transverse wave are perpendicular to its direction of propagation. When one orientation of those vibrations is selected by means of a filter, the wave has been *polarized.* Only transverse waves can be polarized (sound, which is a longitudinal wave, cannot be polarized). Because light is a wave, we can relate the velocity of light to its frequency and wavelength:

$$\mathbf{c} = f\lambda$$

Maxwell's theory, and subsequent experiments, revealed that light is only one form of *electromagnetic radiation*. In Figure 17.1 we see the *electromagnetic spectrum* (not drawn to scale), which displays the full range of electromagnetic waves from high-frequency (short-wavelength) gamma rays (with wavelengths of about 10^{-9} m or shorter) to low-frequency (long-wavelength) radio waves (with wavelengths of about 10 m or longer). In physics, we often express the short wavelengths in units of nanometers (nm) for convenience, but for all calculations, the units should be expressed in the standard unit of meters. The standard frequency unit is still hertz (Hz), but sometimes you will see kilohertz (kHz) and megahertz (MHz) used as well. Again, in all calculations, the standard frequency unit of hertz should be used.

To illustrate how small the wavelengths of visible light are, consider Table 17.1.

To experimentally determine these wavelengths, we will use the properties of interference later in this chapter as we explore Young's double-slit experiment more fully.

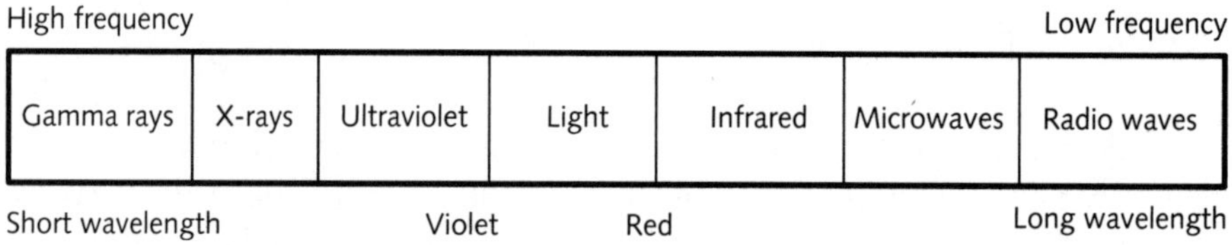

High frequency   Low frequency

Gamma rays	X-rays	Ultraviolet	Light	Infrared	Microwaves	Radio waves

Short wavelength   Violet   Red   Long wavelength

Figure 17.1 The electromagnetic spectrum

Table 17.1   Approximate Wavelengths of Light in a Vacuum

Violet	$4.0 - 4.2 \times 10^{-7}$ m
Blue	$4.2 - 4.9 \times 10^{-7}$ m
Green	$4.9 - 5.7 \times 10^{-7}$ m
Orange	$5.9 - 6.5 \times 10^{-7}$ m
Red	$6.5 - 7.0 \times 10^{-7}$ m

Problem Calculate the frequency of blue light that has a wavelength of 4.5×10^{-7} m.

Solution

Step 1. This is a quick substitution problem. We use the equation:

$$\mathbf{c} = f\lambda$$

and, therefore,

$$f = \mathbf{c}/\lambda = (3 \times 10^{8} \text{ m/s})/(4.5 \times 10^{-7} \text{ m}) = 6.7 \times 10^{14} \text{ Hz}$$

Reflection

Light demonstrates the law of reflection every time you look in a mirror: *the angle of incidence is equal to the angle of reflection.* Both of these angles are measured relative to a normal line drawn perpendicular to a surface (see Figure 17.2).

Refraction

If you have ever looked at a spoon placed into a glass of water from the side, you might have noticed that it looked bent. This is an example of light *refraction.* Your eyes operate using the principles of refraction. However, unlike reflection, refraction occurs when light travels from one transparent substance to another with a different velocity of propagation.

We need to clarify this a bit since if light is incident on a transparent surface along the normal (0°) angle of incidence, then no refraction will occur and the light will pass straight through. In order for refraction to occur, the light must be incident on the new transparent surface at an angle larger than 0° (not along the normal), but not parallel to the surface either. These nonzero angles of incidence are sometimes called *oblique* angles.

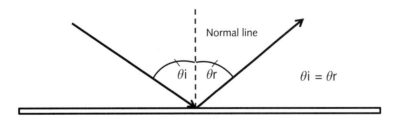

Figure 17.2 Mirror reflection

We can therefore describe the refraction of light (see Figure 17.3) as follows: *refraction is the bending of light as it enters a transparent substance at an oblique angle with a different velocity of propagation.*

Because of the change in velocity of the light (but *not* a change in frequency!), the angle of incidence will *not* be equal to the angle of refraction. If the light is going from air into some other transparent substance, the light will bend *toward* the normal as it slows down. If the light is going from a transparent substance back into the air, then as its velocity increases, it bends *away* from the normal.

We classify the refractive abilities of a transparent material (or medium) by comparing the velocity of light in the medium (**v**) to the velocity of light in air (**c**). This ratio is called the absolute index of refraction (*n*). Notice that, for the rectangular block of glass in Figure 17.3, the light emerges parallel (but offset from) its original angle of incidence. Table 17.2 lists typical absolute indexes of refraction.

The relationship between the angle of incidence and the angle of refraction is known as *Snell's Law* (named for Willebrord Snell, 1580–1626) and is written as follows:

$$n_1 \sin \theta_1 = n_2 \sin \theta_2$$

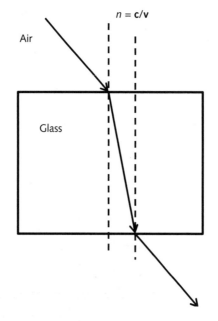

Figure 17.3 Light refraction using a rectangular block of glass

Table 17.2   Absolute Indexes of Refraction

Air	1.00
Corn oil	1.47
Diamond	2.42
Ethyl alcohol	1.36
Glass, crown	1.52
Glass, flint	1.66
Glycerol	1.47
Lucite	1.50
Quartz, fused	1.46
Water	1.33

where n_1 is the absolute index of refraction of the medium where the light is coming *from* and n_2 is the absolute index of refraction of the medium where the light is refracting *to*.

Problem   Light is incident from air onto the surface of Lucite at an angle of incidence equal to 30°. Calculate the angle of refraction and calculate the velocity of light in Lucite.

Solution

Step 1. In the refraction of light from air to Lucite the light is going to be slowing down. We use Snell's Law:

$$(1.00)\sin(30°) = (1.50)\sin \theta_2$$

$$\theta_2 = 19.5°$$

Step 2. To find the velocity of light in Lucite, we use:

$$\mathbf{v} = \mathbf{c}/n = (3 \times 10^8 \text{ m/s})/(1.50) = 2.0 \times 10^8 \text{ m/s}$$

Problem   Light is going from crown glass (n_1 = 1.52) into water (n_2 = 1.33) at an angle of incidence equal to 30°. Calculate the angle of refraction.

Solution

Step 1. The light is going from crown glass (n_1) into water (n_2). We use Snell's Law:

$$(1.52)\sin(30°) = (1.33)\sin \theta_2$$

$$\theta_2 = 34.8°$$

Notice that the angle of refraction is larger than the angle of incidence in this second sample problem. This is because the light is going from a higher-index material (more *optically dense*) to a lower-index material. However, there is a limit to what the angle of refraction can be.

Imagine that light is going from some transparent material into the air. The angle of refraction will be larger because the absolute index of refraction of air is taken to be n_{air} = 1.00. As the angle of incidence increases, the angle of refraction increases (but not linearly!) and remains larger than the angle of incidence. At some *critical angle of incidence* (in the denser material), the angle of refraction will be equal to 90°. If the angle of incidence were to exceed this critical angle of incidence (θ_c), then Snell's Law would no longer be valid (you would get a mathematical error in which the sine of the angle of refraction is greater than 1.0!).

Experimentally, what we observe is that the light actually reflects off the boundary and remains in the original medium (obeying the *law of reflection*). This phenomenon is known as *total internal reflection* and is the basis for modern fiber-optic cables and communications. Total internal reflection can only occur when light goes from a high-index material to a low-index material. For most practical situations, we consider the critical angle of incidence as light goes from some high-index material into the air. At the critical angle of incidence, the angle of refraction is equal to 90° (sin 90° = 1.00) and Snell's Law becomes:

$$\sin \theta_c = 1.00/n_1$$

Problem   Calculate the critical angle of incidence for light going from water into the air.

Solution

Step 1. We use our equation for critical angle from a higher-index material into the air:

$$\sin \theta_c = 1.00/1.33$$

and, therefore,

$$\theta_c = 48.7°$$

For total internal reflection to occur, the angle of incidence of the light in water would have to be greater than this critical angle.

When white light passes through a prism, we observe a continuous spectrum of colors (ROYGBIV). This process is called *dispersion*. When the colors emerge, the red is refracted the least (meaning it travels fastest in the

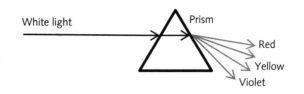

Figure 17.4 Process of dispersion

prism) while the violet is refracted the most (meaning it travels the slowest in the prism). Any material in which the velocity of light is dependent on its frequency is called a *dispersive medium*. With prismatic dispersion, the order of the colors is important. This process is illustrated in Figure 17.4.

Interference and Diffraction

In 1801, Thomas Young demonstrated that light is a wave. In an auditorium filled with people for a public lecture/demonstration, he passed a beam of light through two narrow slits and produced an interference pattern on a screen. The interference was due to the diffraction of light waves through the slits, and (following Huygen's Principle) the slits acted like point sources of interfering waves.

For two or more slits, the interference pattern (for monochromatic light) would appear as a series of bright and dark bands corresponding to the regions of constructive and destructive interference (see Figure 17.5).

If we let x represent the pattern separation distance (between successive maxima), let L represent the distance from the slits to the screen, and let d represent the slit separation distance, then it can be shown that the interference pattern is related to the wavelength of light:

$$\lambda = dx/L.$$

Experimentally, we usually use a *diffraction grating*, which is a 1-inch card that has a clear plastic film with over 13,000 lines scratched on it. In some

Figure 17.5 Multislit diffraction pattern

cases, a typical grating, bought commercially, might have 750 lines/mm in metric units (which means $d = 1.33 \times 10^{-6}$ m). Using a clear, vertical light bulb, a continuous spectrum of colors (ROYGBIV) appears and the wavelengths of light can be determined in a classroom laboratory setting. For a more specific wavelength, a spectrum tube (with hydrogen or other gases) can be used to study the discrete spectral emission lines of these elements (see Chapter 19).

In the case of diffraction, the order of the colors will be *reversed* (VIBGYOR), since the amount of diffraction (x) varies directly with wavelength. This means (since there is no change in velocity) that the violet light diffracts the least (smaller value of x) while the red light diffracts the most (larger value of x). We see this by solving the diffraction equation for x:

$$x = \lambda L/d$$

Consider Young's double-slit experiment. The equation is valid for calculations between the central maximum and the first-order maximum. When the light travels between the slits, the distance from the slits to a point of constructive interference is called the *path length*. For the central maximum, the difference in path length (or *path length difference*) is equal to zero (the central maximum is sometimes called the zero-order maximum). To any other maximum point, the path length difference is equal to a whole multiple of wavelengths ($n\lambda$). Hence, the first-order maximum line has a path length difference equal to one wavelength.

Problem Calculate the wavelength of light when a diffraction grating with 650 lines/mm is used producing an interference pattern 80 cm away and the distance between the central and first-order maximum is 30 cm.

Solution

Step 1. The goal of the problem is to find the wavelength of light, so we first need to find the slit separation distance from the given diffraction grating information:

650 lines/mm → $d = (1/650)$ mm $= 1.54 \times 10^{-3}$ mm $= 1.54 \times 10^{-6}$ m

Step 2. Now we can use the equation for light interference (all units in meters):

$\lambda = (1.54 \times 10^{-6}$ m$)(0.30$ m$)/(0.80$ m$) = 5.78 \times 10^{-7}$ m

Remember, when light slows down, it refracts (bends) toward the normal, and when light speeds up, it refracts (bends) away from the normal. The process of dispersion applies only to refraction, and red light travels faster in a prism than violet light. However, when a continuous spectrum is produced using a diffraction grating, the order of the colors is reversed. This is an important distinction to remember!

As we shall see in Chapter 18, changing the shape of the material will affect the direction of the light as it passes through a transparent material (like a prism or lens shape). Curved mirrors will also produce interesting effects as well and they form the basis of reflecting telescopes and satellite dishes.

Problem-Solving Strategies to Avoid Missteps

When solving problems using the properties of light, it is important to remember that when light refracts, the frequency does not change. The change in velocity is actually accompanied by a change in wavelength. It is possible that you might be asked to construct (using a protractor) reflected or refracted light rays. It is worthwhile to practice this on your own, as we will be concentrating on the calculations. Use the problem-solving ring as needed for guidance.

Exercise 17.1

1. Light is incident from air to the surface of a diamond at an angle of 40°. Calculate the angle of refraction and the velocity of light in the diamond.

2. If the light in question 1 has a frequency of 5.0×10^{14} Hz, calculate the wavelength of the light in air and in the diamond.

3. Light with a wavelength of 5.3×10^{-7} m is incident on a diffraction grating that has 700 lines per mm scratched on it. If a screen is placed 0.8 m away, what is the value of the distance between the central maximum and the first-order maximum lines?

4. Calculate the distance light (in a vacuum) travels in one year (this distance is called a *light-year* in astronomy).

18

Geometrical Optics

In this chapter you will learn about how curved mirrors and lenses form images. These concepts are very important for understanding how the human eye, telescopes, and microscopes operate. We will adapt the laws of reflection and refraction to construct ray diagrams, which will show how the images form. We will not be able to draw these diagrams to scale, but as you practice them, you should have a metric ruler available. The equations related to these concepts are based on certain approximations used by mirror and lens makers.

Images Formed by Curved Mirrors

From our study of waves we can show experimentally that a parabolic shape will cause parallel wave fronts to reflect and converge to a focal point (called a *real focal point*). The distance from the boundary to the focal point is called the *focal length* (f).

If we use a parabolic, *concave* mirror (see Figure 18.1), then the principal axis is an imaginary axis drawn through the mirror that essentially cuts the mirror in half, as shown. A set of light rays parallel to the principal axis will converge at the focal point. The parabolic shape (which is curved) also has a radius of curvature (C). Geometrically, the relationship between the radius of curvature and focal point is given by

$$C = 2f$$

The *law of reflection* still applies, but the shape of the mirror redirects the reflected rays as shown.

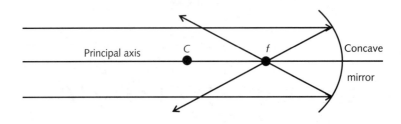

Figure 18.1 Reflection from a concave mirror

If the mirror is convex, then the light rays will diverge away from an imaginary focal point (called a *virtual focal point*) on the other side of the mirror (see Figure 18.2). The image formed by a plane mirror (that appears to be on the other side of the mirror and shows a left-right reversal) is called a *virtual image*.

For the *convex mirror*, light rays that are parallel to the principal axis will reflect *away* from the virtual focal point ($-f$).

 Misconception

It is very important to understand that the ray diagrams use flat mirror approximations. If we draw (or were to actually use) a mirror that was too curved, then the rays would not all focus properly and a condition known as *spherical aberration* would result.

To construct the image formed by a *concave* mirror, we imagine an arrow that represents the object. We define d_o to be equal to the object distance (distance from the object to the mirror) and d_i to be equal to the constructed image distance. The size of the object is represented by S_o and the size of the image is represented by S_i.

We will use two light rays to construct the images. Any light ray parallel to the principal axis of the mirror will reflect through the focal point.

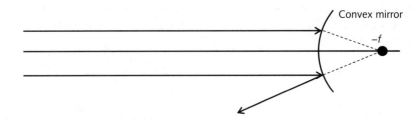

Figure 18.2 Reflection from a convex mirror

Additionally, any light ray passing through the focal point of the mirror will reflect parallel to the principal axis. We will construct five cases using concave (converging) mirrors (see Table 18.1).

Each of these cases is summarized in a not-to-scale sketch in Figures 18.3–18.7. The actual values of points f and C are not important for these sketches. In an actual construction, you would need to use a metric

Table 18.1 Images Formed by a Concave Mirror

CASE	IMAGE
I	Object beyond point C ($d_o > 2f$); see Figure 18.3
II	Object at point C ($d_o = C = 2f$); see Figure 18.4
III	Object between points C and f ($C < d_o < f$); see Figure 18.5
IV	Object at f ($d_o = f$); see Figure 18.6
V	Object between f and the mirror ($d_o < f$); see Figure 18.7

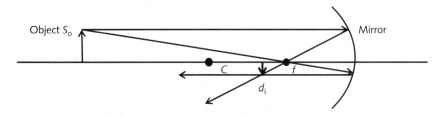

Figure 18.3 Case I (object beyond point C): Image is real, inverted, smaller, and located between points f and C

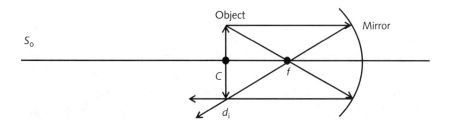

Figure 18.4 Case II (object at point C): Image is real, inverted, same size, and located at point C

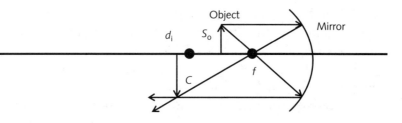

Figure 18.5 Case III (object between points *f* and *C*): Image is real, inverted, larger, and located beyond point *C*

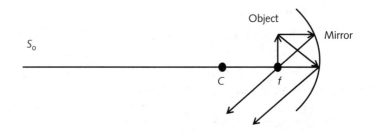

Figure 18.6 Case IV (object located at focal point): No image is formed (light ray incident at center of mirror follows simple reflection)

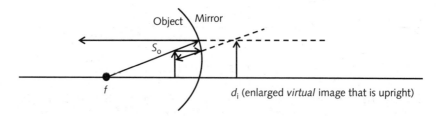

Figure 18.7 Case V (object between *f* and the mirror): Enlarged, virtual, upright image

ruler to make as accurate a construction as possible. When the light rays actually cross each other, we say that a *real image* is formed (real images are also always inverted!). These real images can also be projected onto a screen.

Note that for case V, the image formed could not be projected since it is not *real* and appears to be *behind* the mirror. It will be important to remember that real images are *always inverted* and will be algebraically *positive*. Virtual images are always *upright* and will be algebraically *negative*. Real focal points are algebraically *positive* (for concave mirrors). Notice the light

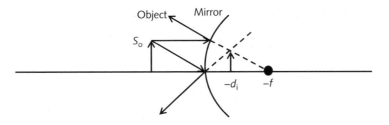

Figure 18.8 The image formed is smaller, virtual, upright, and located behind the mirror ($d_i < 0$)

rays drawn to produce the image in case V. You should memorize these cases and their consequences.

The image formed by a convex (diverging) mirror involves only one case because there is no real focal point. The ray diagram is shown in Figure 18.8 (the focal length is considered negative!).

Notice that to form the virtual image, we again used a light ray directed toward the center of the mirror that follows the law of reflection.

The relationship (for a flat mirror approximation) between the object distance (d_o), image distance (d_i), and focal length (f) is given by the equation:

$$\frac{1}{f} = \frac{1}{d_o} + \frac{1}{d_i}$$

For the object size (S_o) and the image size (S_i), the equation relating them is:

$$\frac{S_i}{S_o} = \left|\frac{d_i}{d_o}\right|$$

Problem A 2-cm-tall object is located 10 cm in front of a concave mirror that has a radius of curvature equal to 8 cm. Calculate the image distance and the image size.

Solution

Step 1. First we need to find the focal length of the mirror. Since $C = 2f$, we know that $f = 4$ cm and this is a case I situation (image should be smaller and located between f and C).

Step 2. We substitute in our given values with d_o = 10 cm and f = 4 cm:

$$\frac{1}{4}\ \text{cm} = \frac{1}{10}\ \text{cm} + \frac{1}{d_i}$$

d_i = 6.67 cm (the positive answer means the image is real)

Step 3. We can easily find the size of the image:

$$S_i = (\frac{d_i}{d_o})S_o = \left(6.67\ \text{cm}/10\ \text{cm}\right)\left(2\ \text{cm}\right) = 1.33\ \text{cm}$$

Try doing the sample problem using a scaled ray diagram and see how your results compare.

Concave mirrors are used in reflecting telescopes. If you look at photographs of radio telescopes (and dish antennae for satellite television), you will see the same basic parabolic shapes.

Images Formed by Lenses

When light is refracted, its direction is changed due to the change in the velocity of light and the angle of incidence (recall Chapter 17). Changing the shape of the refracting material can produce some interesting effects.

A typical double convex lens (which produces real images) is shown in Figure 18.9. Due to the difference in the absolute index of refraction, different materials will cause the light to focus in different places (higher refractive indices will have shorter focal lengths). Another problem encountered with convex lenses is the dispersive quality of glass, which might create *chromatic aberration*, a distortion of colors that causes the different colors to focus at different places along the principal axis. In optics, we assume that the lens is thin so that we can more easily construct the ray diagrams. Interestingly, the cases for convex lenses (as well as the equations we use) will be similar to the ones used for concave mirrors. Convex lenses, often

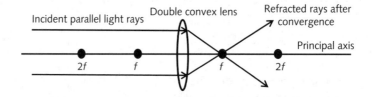

Figure 18.9 Refraction through a double convex lens

referred to as *converging lenses*, are used in telescopes, microscopes, projectors, and the human eye.

Since the lens in Figure 18.9 is a double convex lens, there is a symmetry that produces the two *real* focal points. Instead of a center of curvature (*C*), we label the point *2f*. From the diagram we see that all incident light rays that are parallel to each other refract through the focal point. Additionally, any incident light ray passing through the focal point will refract parallel to the principal axis. Convex lenses can produce real images that can be projected onto a screen.

Concave lenses produce small virtual images (see Figure 18.10). They are referred to as *diverging* lenses because parallel light rays that are incident on the lens refract away from an imaginary (*virtual*) focal point. Algebraically, we will treat the focal lengths for concave lenses as *negative*. We again use a thin lens approximation when making the ray diagrams.

Misconceptions

It is very important not only to memorize all of the cases but to clearly distinguish between mirrors and lenses. As you will see, it is the convex lenses that produce the real images using refraction, but it is the concave mirrors that produce real images using reflection. Also, you must be careful with the signs; real images for curved mirrors form on the same side as the objects while for convex lenses, the real images form on the other side of the lenses!

We will construct five cases for convex lenses (Figures 18.11–18.15). They are very similar to the five cases for concave mirrors except we use the point *2f* instead of the center of curvature *C* (see Table 18.2). Remember that lenses *refract* and mirrors *reflect*! Also, a light ray through the center of the lens will not be refracted in our thin lens approximation. The equations will be the same as before with the curved mirrors.

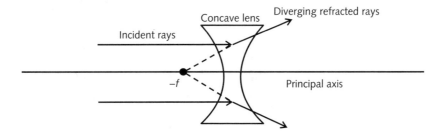

Figure 18.10 Refraction through a double concave lens

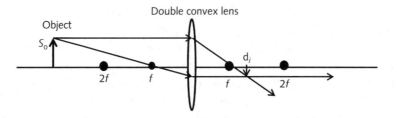

Figure 18.11 Case I (object beyond 2*f*)

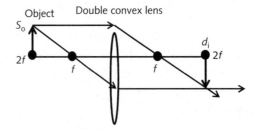

Figure 18.12 Case II (object at 2*f*): Image is real, inverted, same size, and located at 2f object

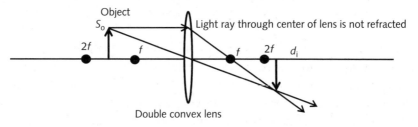

Figure 18.13 Case III (object between *f* and 2*f*): Image is real, inverted, larger, and beyond 2*f*

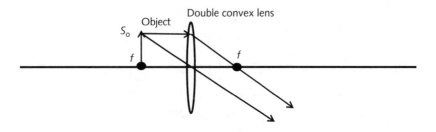

Figure 18.14 Case IV (object at *f*): No image is formed!

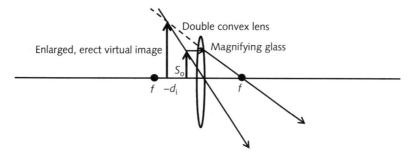

Figure 18.15 Case V (object between f and the lens): Enlarged, virtual, upright image

Table 18.2   Images Formed by a Convex Lens

CASE	IMAGE
I	Object beyond point $2f$ ($d_o > 2f$); see Figure 18.11
II	Object at point $2f$ ($d_o = 2f$); see Figure 18.12
III	Object between points $2f$ and f ($2f < d_o < f$); see Figure 18.13
IV	Object at f ($d_o = f$); see Figure 18.14
V	Object between f and the lens ($d_o < f$); see Figure 18.15

You will notice that the first three cases produce *real*, inverted images and the last case produces an enlarged *virtual* image (acts like a magnifying glass). The virtual image is upright and the image distance is taken as *negative*. Notice that for the convex lens, the virtual image is appearing on the same side as the object (compare with the concave mirror cases). For the concave lens, only one case is needed and shown (Figure 18.16). A small upright virtual image is *always* produced with this type of lens. The focal length for a concave lens is *negative*.

The equations used for lenses are the same as the equations used for mirrors.

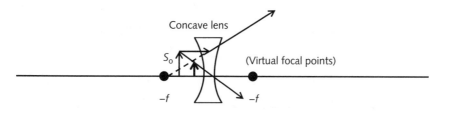

Figure 18.16 Concave lens construction

Problem A 2-cm-tall object is placed 10 cm in front of a concave lens that has a focal length of –20 cm. Calculate the image distance and the image size.

Solution

Step 1. The equations for curved mirrors and lenses are the same. A concave lens has a negative focal length. Therefore:

$$\frac{1}{f} = \frac{1}{d_o} + \frac{1}{d_i}$$

and

$$\frac{S_i}{S_o} = \left| \frac{d_i}{d_o} \right|$$

Step 2. $\dfrac{1}{-20} \text{ cm} = \dfrac{1}{10} \text{ cm} + \dfrac{1}{d_i}$

$d_i = -6.67$ cm (a virtual image)

Step 3. $S_i = |(-6.67 \text{ cm})/(10 \text{ cm})|(2 \text{ cm}) = 1.33$ cm (image size is smaller)

Problem-Solving Strategies to Avoid Missteps

The key to solving ray diagrams is to remember all of the cases and the appropriate rays to construct for mirrors and lenses. None of the sample constructions shown in this book were drawn to scale, but you could easily reproduce the same problems as ray diagrams and compare your results. Templates will be shown in some of the end-of-chapter exercises, but it is always important to practice. Make sure you understand the differences between real and virtual images.

Exercise 18.1

1. A convex lens forms an image (on the right) of an object (on the left) as shown in the diagram. Use a ray diagram to locate the symmetrical focal points on each side of the lens. A ruler is not necessary.

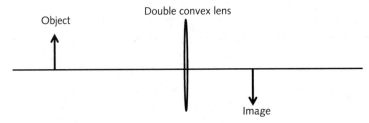

Double convex lens

Object

Image

2. A 2-cm-tall object is located 8 cm in front of a concave mirror that has a radius of curvature equal to 10 cm. Calculate the location of the image and its size.

3. On the diagram shown, construct the ray diagram to locate the image of the object shown in the form on a convex mirror. A ruler is not necessary.

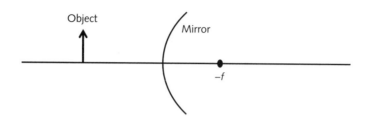

4. An object is located 8 cm in front of a convex lens. A real image forms 12 cm on the other side of the lens. Calculate the focal length of the lens.

5. Where should a light bulb be placed relative to a concave mirror so that the reflected light rays are parallel to each other?

19

Quantum Theory of Light

In this chapter you will learn about the dual nature of light. Thomas Young demonstrated the wave nature of light using interference and diffraction through multiple slits. In the nineteenth century the wave theory was theoretically established by James Clerk Maxwell and experimentally studied by Heinrich Hertz. In the twentieth century Max Planck and Albert Einstein began the process of studying the quantum theory of radiation, and Robert Millikan (in addition to his famous oil drop experiment for electron charge) verified the photoelectric effect and the existence of photons. This led to the discovery of matter waves and photon momentum. Matter and energy would be forever linked through Einstein's equation $E = mc^2$ and his theory of special relativity. This is the beginning of the final steps in our journey of exploring introductory physics.

Photoelectric Effect

In 1900, Max Planck (1858–1947) proposed that electromagnetic radiation could be explained theoretically by postulating the existence of discrete packets of energy called *quanta* (or photons). He described the energy of a photon as directly proportional to the corresponding frequency of light and given by the equation

$$E = hf = hc/\lambda \text{ (since } \mathbf{c} = f\lambda)$$

where $h = 6.63 \times 10^{-34} \text{ J} \cdot \text{s}$, which is known as *Planck's constant* (recall that $\mathbf{c} = 3 \times 10^8 \text{ m/s}$).

This was a new way of looking at radiation. Planck was interested in understanding the relationship between the observed color of light emitted when objects are heated (think of the colors of stars). As we shall see, Planck's theory was going to be in direct conflict with Maxwell's theory of electromagnetic radiation, which stated that the oscillation of electromagnetic fields produced energy that traveled in the form of transverse waves and propagated with the velocity of light in a vacuum.

Problem Calculate the energy of a photon (in joules and electron-volts) associated with light that has a frequency of 6.5×10^{14} Hz.

Solution

Step 1. This is a simple substitution problem and we begin by using Planck's equation:

$$E = hf = (6.63 \times 10^{-34} \text{ J} \cdot \text{s})(6.5 \times 10^{14} \text{ Hz}) = 4.3 \times 10^{-19} \text{ J}$$

Step 2. To convert to electron-volts (eV), we simply recall that
$$1 \text{ eV} = 1.6 \times 10^{-19} \text{ J}$$

$$E = 2.69 \text{ eV}$$

During his short life, Heinrich Hertz (1857–1894) verified Maxwell's theoretical predictions experimentally. He noticed a curious effect: certain metals seemed to emit electrons when certain frequencies of radiation were incident on them. This phenomenon became known as the photoelectric effect. For example, if a piece of zinc is placed on an electroscope and then negatively charged, the electroscope will be discharged by ultraviolet light. However, if the electroscope and zinc are positively charged, the effect does not occur.

In 1905, in addition to developing his paper on the special theory of relativity (which we will not discuss in this book), Albert Einstein (1879–1955) explained the photoelectric effect using Planck's quantum theory of light. Einstein proposed that the free surface electrons in a metal can be released if they are given a minimum amount of energy called the *work function*. This work function (designated in some books as W_o or φ) depends on a minimum frequency of radiation (f_o) called the *threshold frequency*. Using Planck's theory, Einstein showed that

$$W_o = hf_o$$

If electromagnetic radiation (in the form of a photon with frequency f) is incident on a metal and its frequency is above the threshold frequency,

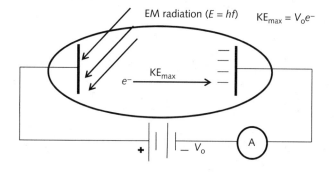

Figure 19.1 Typical photoelectric effect experiment

then a surface electron can absorb the energy and be released with excess kinetic energy:

$$KE = hf - hf_o$$

Robert Millikan (1863–1958) experimentally verified Einstein's theory (see Figure 19.1) using metals in a vacuum. When electromagnetic radiation was incident on the metal, the electron emitted could be detected at the anode, which was part of a complete circuit. An ammeter would record the *photocurrent* (the rate of electron emission).

Millikan verified that the rate of electron emission was dependent on the intensity of the radiation once it is above the threshold frequency. To measure the kinetic energy of the electrons, a variable *negative* potential difference was applied to the anode and adjusted until the current was stopped. This stopping potential (or stopping voltage V_o) is directly related to the kinetic energy of the electrons and was found to be *independent* of the intensity of the incident radiation! This was a clear contradiction of Maxwell's theory. The emitted electrons had a maximum kinetic energy that depended only on the work function of the metal and the frequency of the incident radiation. Einstein's photoelectric equation is now rewritten as:

$$KE_{max} = hf - hf_o = E_{ph} - W_o$$

and $KE_{max} = V_o e^-$, where e^- is equal to the *magnitude* of the electron's charge ($1.6 \times 10^{-19}C$).

Millikan graphed the maximum kinetic energy of emitted electrons against the frequency of incident radiation for many metals (see Figure 19.2) and found that all the graphs had the same slope even if they had different threshold frequencies. The slope was equal to Planck's

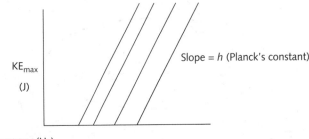

Figure 19.2 Graph of emitted electrons vs. frequency of incident radiation (metals)

constant h (as predicted by Einstein). The photoelectric effect verifies the photon nature of light, and Young's double slit experiment verifies the wave nature of light!

Problem   Light with a frequency of 1.5×10^{15} Hz is incident on a metal that has a threshold frequency of 5.7×10^{14} Hz. Calculate the work function of the metal (in joules and eV), the maximum kinetic energy of the emitted electrons (in joules and eV), and the stopping potential.

Solution

Step 1. We can see that the incident frequency exceeds the threshold frequency. Therefore, the photoelectric effect is valid. To find the work function we use:

$$W_o = hf_o = (6.63 \times 10^{-34} \text{ J} \cdot \text{s})(5.7 \times 10^{14} \text{ Hz}) = 3.78 \times 10^{-19} \text{ J} = 2.36 \text{ eV}$$

Step 2. The maximum kinetic energy is given by:

$$KE_{max} = hf - W_o = (6.63 \times 10^{-34} \text{ J} \cdot \text{s})(1.5 \times 10^{15} \text{ Hz}) - 3.78 \times 10^{-19} \text{ J}$$

$$KE_{max} = 6.17 \times 10^{-19} \text{ J} = 3.85 \text{ eV}$$

Notice that the energy of the incident photons is equal to 9.95×10^{-19} J.

Step 3. By having the maximum kinetic energy (which is the energy of one emitted electron as well as one million emitted electrons) expressed in electron-volt units, we immediately see that:

$$KE_{max} = 3.85 \text{ eV}$$

Then (by definition)

$$V_o = 3.85 \text{ V}$$

Problem In a photoelectric effect experiment, electrons are emitted and a stopping potential of 2.8 V is needed to measure their kinetic energy. What is the value of the maximum kinetic energy of each electron? If the material used has a work function 0.85 eV, what is the value of the threshold frequency? What is the energy of the incident electromagnetic radiation and what is its frequency?

Solution

Step 1. Since the stopping potential in the experiment is equal to 2.8 V, this means that the maximum kinetic energy for each electron must be equal to 2.8 eV. Thus,

$$KE_{max} = 2.8 \text{ eV} = 4.48 \times 10^{-19} \text{ J (recall that 1 eV} = 1.6 \times 10^{-19} \text{ J)}$$

Step 2. We are given that $W_o = 0.85$ eV; this means that the energy of the incident photons is equal to:

$$E_{ph} = KE_{max} + W_o = 3.65 \text{ eV} = 5.84 \times 10^{-19} \text{ J}$$

Step 3. We use Planck's formula to find the frequency (make sure the energy is in joules!):

$$E_{ph} = hf$$

and, therefore,

$$f = (5.84 \times 10^{-19} \text{ J})/(6.63 \times 10^{-34} \text{ J} \cdot \text{s}) = 8.81 \times 10^{14} \text{ Hz}$$

Incidentally, the threshold frequency for this material would be equal to $f_o = 2.05 \times 10^{14}$ Hz.

Misconceptions

The fact that light has a dual nature can be very abstract and confusing. It is important to remember that we base our understanding of light on experimental observations. If a ray of light is incident on two slits, it will produce a characteristic *wave* interference pattern. However, the photoelectric effect can *only* be explained in terms of the *quantum theory of light* and the hypothesis that light consists of packets of energy called photons. Photons do not have any *rest mass*, but they travel with the velocity **c** in a vacuum and do possess *momentum*!

The photoelectric effect demonstrates that the maximum kinetic energy of the emitted electrons is independent of the intensity of the light (not the amplitude of a wave, but the number of incident photons!) and directly proportional to the frequency of the incident light. Increasing the intensity of the light will increase the current (rate of electron emission), but *will not* change the maximum kinetic energy. These are crucial ideas that form a contrast between what we call *modern physics* and our previous work in *classical physics*.

Matter Waves

If light can behave like a particle, can a particle behave like a wave? This may seem like an absurd question, but the ideas in quantum theory are very strange indeed. Louis deBroglie (1892–1987) developed a theory of matter waves that was to influence the study of the atom in the 1920s. DeBroglie suggested (and this was later confirmed by experiments) that particles moving with velocities less than the velocity of light could have a wavelike nature. His theory predicted the existence of *matter waves* such that the wavelength of a matter wave is:

$$\lambda = h/mv \text{ (where } h \text{ is equal to Planck's constant)}$$

Experiments show that when a beam of electrons is sent through two narrow slits, a wavelike diffraction pattern appears. This effect is observable only because of the small mass of the electron (9.1×10^{-31} kg). We could *not* observe the matter wavelength for a 1-kg mass! These concepts have been put to practical use with the invention of the scanning electron microscope.

Problem Calculate the matter wavelength for an electron that has a velocity of 2×10^7 m/s. Compare this to the matter wavelength for a 1-kg block moving at the same velocity.

Solution

Step 1. We calculate each matter wavelength using the deBroglie equation:

$\lambda_{electron}$ = (6.63 × 10^{-34} J·s)/(9.1 × 10^{-31} kg)(2 × 10^7 m/s) = 3.64 × 10^{-11} m (observable)

λ_{block} = (6.63 × 10^{-34} J·s)/(1 kg)(2 × 10^7 m/s) = 3.32 × 10^{-41} m (not observable)

Compton Effect and Photon Momentum

Albert Einstein predicted, using his theory of special relativity, that matter and energy were related through his famous equation $E = m\mathbf{c}^2$ (where $\mathbf{c}$ is equal to the velocity of light). Thus, a 1-kg mass theoretically has a self-energy equal to 9×10^{16} J (an enormous amount of energy). It is not possible to extract that energy from ordinary matter, but as we will learn, nature does it all the time on the atomic and nuclear scale. Since Planck demonstrated that for a photon $E = hf$, what happens if we mix Planck's equation with Einstein's equation?

$$E = m\mathbf{c}^2 = hf = h\mathbf{c}/\lambda \quad \text{(in free space } \mathbf{c} = f\lambda)$$

Arthur Compton (1892–1962) was able to demonstrate that in certain circumstances a photon can collide with an electron and conserve both energy and momentum. This is known as the *Compton effect*. Even though they do not possess any *rest mass*, photons do have a momentum ($\mathbf{p}$) that varies inversely with their wavelength:

$$\mathbf{p} = h/\lambda$$

High-energy gamma rays, for example, have a very high momentum when compared with microwaves. Thus both light and matter have a dual nature. Under the right conditions, particles can behave like waves and waves can behave like particles!

Problem Calculate the momentum of a photon that has a frequency of 3.5×10^{15} Hz.

Solution

Step 1. To use the photon momentum equation we need to calculate the wavelength:

$$\lambda = \mathbf{c}/f = (3 \times 10^8 \text{ m/s})/(3.5 \times 10^{15} \text{ Hz}) = 8.57 \times 10^{-8} \text{ m}$$

Step 2. We can now calculate the photon momentum:

$$\mathbf{p} = h/\lambda = (6.63 \times 10^{-34} \text{ J} \cdot \text{s})/(8.57 \times 10^{-8} \text{ m}) = 7.74 \times 10^{-27} \text{ kg} \cdot \text{m/s}$$

This photon momentum may not seem like a large number on our scale, but on the atomic or nuclear scale, it can have a significant impact. Scaling will be important as we study our last topic, atomic and nuclear physics.

Problem-Solving Strategies to Avoid Missteps

We have already outlined the misconceptions associated with the quantum theory of light in general and the photoelectric effect in particular. It is always important to keep track of units (joules vs. electron-volts) and carefully examine each question for the given information and the solutions that need to be solved for. As always, organize your problem-solving path and be careful!

Exercise 19.1

1. Electromagnetic radiation with a frequency of 4.5×10^{14} Hz is incident on a metal that has a threshold frequency of 1.5×10^{14} Hz. Calculate the energy of the incident photons in J and eV. Calculate the work function of the metal in J and eV. Calculate the maximum kinetic energy of the emitted electrons in J and eV.

2. It is observed that when light with a frequency of 5.8×10^{14} Hz is incident on a metal, 1.5 V of negative potential difference are needed to stop electron emission. Calculate the work function for this metal.

3. Calculate the energy (in joules) associated with photons of light that have a wavelength of 565 nm.

4. Protons have a mass of 1.67×10^{-27} kg. Calculate the deBroglie matter wavelength of the protons if they have a velocity of 1.5×10^8 m/s.

5. A photon scattering experiment determines that incident photons have a momentum of 7.5×10^{-25} kg·m/s. Calculate the wavelength associated with these photons.

20

Atomic and Nuclear Physics

In this chapter we will finish our exploration of introductory physics by looking at the structure of the atom and the nucleus. There are many new and exciting ideas probing the very fundamentals of matter (such as subatomic particles and quarks). As we bring our journey to a close, think about how we have progressed from the idea of measurements and forces in classical physics, to the strange world of quantum mechanics, where probability and chance govern interactions.

Spectral Lines

The idea of matter being made up of small, fundamental particles goes back to ancient Greece. Democritus (around 460 BC) was one of the first to postulate the existence of atoms. At the beginning of the nineteenth century, chemist John Dalton (1766–1844) developed his *law of multiple proportions* in which he proposed that chemical compounds were made up of discrete combinations of fundamental building blocks of matter of definite, but different, weights. In 1827, botanist Robert Brown (1773–1858) discovered the random motion of pollen grains suspended in water. This motion (now known as *Brownian motion*) is due to the collision of microscopic water molecules in constant motion. Finally, chemists began to break matter apart into smaller constituent elements. Dmitri Mendeleev (1834–1907) was the first to organize these elements into a period table and predict the existence of new elements according to the arrangement of fundamental weights and electric charges (later known as the atomic mass and atomic number).

In a different area of physics and chemistry, scientists were discovering that when certain chemical elements were heated, they gave off a characteristic color of light. When this light was examined by a spectroscope (a device consisting of either a prism or a diffraction grating), they discovered that what appeared to be a continuous distribution of colors (like the continuous light emitted from an incandescent source) was in fact a series of discrete colored lines. Each element had its own characteristic emission spectrum that could be used in chemical analysis.

A physical mechanism that could explain the production of these lines of color was not available until the early twentieth century. In the meantime, astronomers discovered *absorption lines* in the spectrum of the sun and the stars. The element helium was discovered this way (the name is derived from the ancient Greek word *helios*, referring to the sun god). The absorption lines produced by the stars were in layers deeper and hotter than the outer layers observed (called photospheres) and appeared dark against a continuous colored background but in approximately the same location as the corresponding emission lines of the elements. For example, astronomers observed that the sun contained spectral lines associated with hydrogen, helium, calcium, and iron. The color of a star was related to its temperature and chemical composition.

At the end of the nineteenth century, Johann Jakob Balmer (1825–1898) developed a mathematical formula that related the visible lines in the hydrogen spectrum to each other (hydrogen is the simplest and most abundant element in the universe). The visible hydrogen lines are now called *Balmer lines* (see Figure 20.1). Balmer's formula was motivated by the fact that the visible hydrogen spectrum reaches a limit at a wavelength of 3.645×10^{-7} m (346.5 nm):

$$\lambda \, (\mathrm{m}) = \frac{(3.645 \times 10^{-7} \, \mathrm{m})n^2}{n^2 - 4}$$

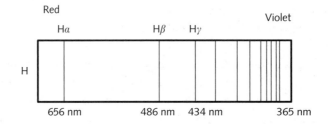

Figure 20.1 Visible hydrogen spectrum

where

$$n = 3, 4, 5, \ldots$$

Balmer could predict the existence of new visible hydrogen lines, but the meaning of the index n was unknown. In the early twentieth century, the ultraviolet (UV) hydrogen spectrum was discovered by Theodore Lyman (1874–1954) and the infrared (IR) hydrogen spectrum by Friedrich Paschen (1865–1947). Each of the different sets of spectra could be written in a similar form that was developed by Johannes Rydberg (1854–1919). For the Balmer series he found:

$$\frac{1}{\lambda} = R\left(\frac{1}{2^2} - \frac{1}{n^2}\right)$$

$R = 1.0968 \times 10^7 \text{ m}^{-1}$ (Rydberg constant)   and $n = 3, 4, 5, \ldots$

For the Lyman UV series, the equation would be:

$$\frac{1}{\lambda} = R\left(\frac{1}{1^2} - \frac{1}{n^2}\right)$$

where

$$n = 2, 3, 4, \ldots$$

For the Paschen IR series, the equation would be:

$$\frac{1}{\lambda} = R\left(\frac{1}{3^2} - \frac{1}{n^2}\right)$$

where

$$n = 4, 5, 6, \ldots$$

The factor of $1/\lambda$ is *proportional* to the energy of the emitted photons since $E = hf = hc/\lambda$.

Problem   Calculate the wavelength for the fourth visible hydrogen line (Hδ) using the Rydberg formula and $n = 6$. Recall that 1 nm = 1×10^{-9} m.

Solution

Step 1. We substitute $n = 6$ into the Balmer series version of the Rydberg formula:

$$\frac{1}{\lambda} = R\left(\frac{1}{2^2} - \frac{1}{6^2}\right) = 2.438 \times 10^6 \ \mathrm{m^{-1}} \rightarrow \lambda = 4.10 \times 10^{-7} \mathrm{m} \ (410 \ \mathrm{nm})$$

Radioactive Decay

In 1896 Henri Becquerel (1852–1908) discovered that a rock made of potassium uranyl sulfate emitted a mysterious form of energy that could expose sealed photographic paper. He realized the uranium was responsible for these emanations. In the next few years, Marie Curie (1867–1934) and her husband Pierre Curie (1859–1906) discovered two new elements (named polonium and radium) that were strong emitters of what they called *radioactivity*.

Around 1903, Ernest Rutherford (1871–1937) along with Fredrick Soddy (1877–1956) discovered that the emissions produced by radioactive elements consisted of three parts. Using magnetic fields, they discovered that one set of rays (called *alpha rays*) were actually positively charged helium nuclei. The second set of rays (called *beta rays*) were the small, negatively charged particles identified as electrons. The electron was discovered in 1897 by J. J. Thomson (1856–1940) using cathode ray tubes. The final set of rays (called *gamma rays*) were high-energy photons; these were not affected by electromagnetic fields. Radioactive elements decayed into new elements by a process known as *transmutation*. We will discuss this process later in the chapter.

Rutherford discovered that the rate of emission activity diminished over time (exponentially) and was proportional to a unit of time called the *half-life*. The half-life of a radioactive element is the time that it takes for half of a given amount of radioactive material to decay. Some elements have a half-life measured in milliseconds; others can have a half-life of over 1 billion years! If n represents the number of half-lives, then the radioactive decay of the mass is given by:

$$m_f = \frac{m_i}{2^n}$$

Problem If a radioactive material has a half-life of 24 min, how many grams will remain after 144 min if we began with a sample of 100 g?

Solution

Step 1. We first need to know the number of half-lives:

$$n = (144 \text{ min})/(24 \text{ min}) = 6$$

Step 2. The remaining mass is therefore equal to:

$$m_f = (100 \text{ g})/(2^6) = 1.56 \text{ g}$$

Bohr Model of the Atom

When J. J. Thomson discovered the electron, he proposed a model of the atom that needed to be electrically neutral. He imagined the atom as a large positive charge with scattered negative electrons inside. This model became known (in one version) as the *raisin bun* model.

There were some theoretical problems with Thomson's model, and Rutherford, along with his colleagues, decided to use alpha particles to probe the atom. In 1911 Rutherford published the results of his experiments in which he fired alpha particles at a thin sheet of gold foil. While most of the alpha particles passed through the gold foil (as detected by a scintillation screen), some were deflected into hyperbolic paths.

Rutherford analyzed his results theoretically and concluded that the atom consisted of a very small positively charged center (which he called the *nucleus*) surrounded by negatively charged electrons. In Rutherford's model, the electrons orbited the nucleus in a way similar to the way planets orbit the sun. This idea was prompted by the fact that electrostatic attraction (Coulomb's Law) was the main force between the electrons and the nucleus. The atom was mostly empty space.

However, there were also theoretical problems with Rutherford's ideas. Classical physics predicted that accelerated charges would produce electromagnetic radiation. An orbiting electron has a centripetal acceleration and so should produce continuous radiation (which it does not). Furthermore, the continuous emission of energy would quickly cause the electrons to spiral into the nucleus (which, again, does not happen). A new and radical approach to the atom was needed.

In 1913 Niels Bohr (1885–1962) developed a theory based on the quantum theory of light developed by Planck and Einstein. The index n that appeared in the spectral line equations of Balmer and Rydberg was now recognized as being related to discrete energy states. Bohr proposed that despite the electron's orbital motion, it did not emit electromagnetic radiation. Instead, the electrons existed in energy (or quantum) states, the lowest energy state being called the *ground state*. For the hydrogen atom, Bohr was able to reproduce the Rydberg formula and accurately predict all of the visible Balmer lines. However, Bohr's theory did not explain more complex atoms or molecules.

In essence, Bohr's theory calculated the energy states (later verified by experiment) for the hydrogen atom and stated that the hydrogen atom would emit visible light photons when electrons "fall" from higher energy states to level 2 (recall the Balmer formula):

$$E_{ph} = E_i - E_f = hf$$

A transition to a higher energy state would produce an *absorption* line. The energy of the emitted photons would be equal to the exact energy difference between states (visible lines to level 2; ultraviolet lines to level 1; infrared lines to level 3). This is consistent with the Rydberg spectral formula.

An electron in a given energy state can absorb a photon only if the photon is equal to the energy difference between states. An atom can become ionized if an electron interacts with a photon that has more energy than the energy difference between that state and infinity (see Figure 20.2). For example, an electron in the ground state will become ionized if it absorbs a photon with an energy greater than 13.6 eV. However, if a photon interacts with the electron (in the ground state) with energy less than 10.2 eV (the difference between energy states 1 and 2), the electron will not change states.

Problem   Calculate the frequency and wavelength of a photon emitted in a hydrogen atom as electrons fall from level 2 to level 1 (referred to as the Lyman alpha line, Lyα).

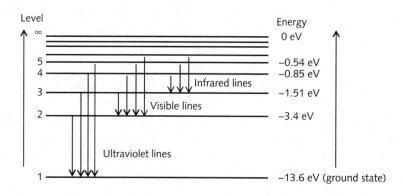

Figure 20.2 Hydrogen energy states

Solution

Step 1. We need the energy difference $E_2 - E_1 = (-3.4 \text{ eV}) - (-13.6 \text{ eV}) = 10.2 \text{ eV}$.

Step 2. We convert this energy difference to joules and then use the Planck equation:

$$E_{ph} = (10.2 \text{ eV})(1.6 \times 10^{-19} \text{ J/eV}) = 1.632 \times 10^{-18} \text{ J} = (6.63 \times 10^{-34} \text{ J} \cdot \text{s})f$$

$$f = 2.46 \times 10^{15} \text{ Hz}$$

Step 3. To find the wavelength, we use the fact that $\mathbf{c} = f\lambda$:

$$\lambda = \mathbf{c}/f = (3 \times 10^8 \text{ m/s})/(2.46 \times 10^{15} \text{ Hz}) = 1.22 \times 10^{-7} \text{ m} = 122 \text{ nm (UV)}$$

In the 1920s the use of deBroglie matter waves led Erwin Schrodinger, Niels Bohr, Werner Heisenberg, and others to develop the theory of *quantum mechanics*. According to this theory, interactions on an atomic and subatomic scale are governed by probability and chance. The electrons have indefinite positions (represented by an *electron cloud*), with the concept of uncertainty (as embodied in Heisenberg's *uncertainty principle*) implying that we can never know for certain both their position and momentum (the electron is actually represented by an abstract mathematical entity called a *wave function*).

Quantum mechanics is a remarkable, but very strange, theory. It has proven to be very successful in explaining much about the structure of the atom and elementary particles, but it contradicts the classical ideas of gravity as described by Einstein's theory of general relativity (which treats the gravitational force as a manifestation of a warping of space-time due to the presence of mass). Nonetheless, quantum mechanics has been unified with special relativity, and the interactions between photons and electrons (called *quantum electrodynamics* or *quantum field theory*) have led to deeper insight into the subatomic world. Antimatter was predicted and then discovered, Einstein's famous equation $E = mc^2$ was verified, and his theory of relativity plays a very important role in nuclear energy production.

As of today, physicists are trying to unite all of the laws of physics (especially general relativity and quantum mechanics) with abstract concepts such as strings, membranes, and a universe with 11 dimensions! We have come a long way on our step-by-step journey that began with the uncertainty of simple measurements of length on a scale of a few centimeters or meters and ends with some exciting frontiers in the twenty-first century that look at the universe on a scale of 10^{-33} m.

Misconceptions

It is very important to understand that the Bohr model is appropriate for hydrogen and hydrogen-like atoms only. Sometimes you may see energy-level diagrams given for mercury. It is not realistic to think of an atom as really consisting of orbiting point particles even though that is how we are treating them. You must also remember that to use the equations given in standard units, the energy-level differences must be in units of joules.

Nuclear Models

The atomic model of Bohr and the nuclear investigations of Rutherford led to some familiar concepts you may have learned in chemistry. The nucleus consists of protons (which are positively charged) and neutrons (which are electrically neutral).

Each atomic element is expressed in terms of its charge (atomic number) and mass number (number of protons and neutrons in its nucleus). Nuclei with the same number of protons but a different number of neutrons are called isotopes. For example:

$$\text{Hydrogen} = {}_{1}^{1}\text{H}\left(\text{proton}\right)$$

$$\text{Deuterium} = {}_{1}^{2}\text{H}$$

$$\text{Tritium} = {}_{1}^{3}\text{H}$$

$$\text{Helium} = {}_{2}^{4}\text{He}\left(\text{alpha particle}\right)$$

$$\text{Lithium} = {}_{3}^{7}\text{Li}$$

$$\text{Carbon} = {}_{6}^{12}\text{C}$$

$$\text{Electron} = {}_{-1}^{0}\text{e}$$

$$\text{Neutron} = {}_{0}^{1}\text{n}$$

$$\text{Gamma ray photon} = {}_{0}^{0}\gamma$$

Using magnetic fields and mass spectrometers, physicists can measure the actual masses of nuclei. A universal mass unit is defined in nuclear physics such that:

1 u = 1/12 mass of a carbon-12 nucleus (approximately 1.6605×10^{-27} kg)

In these units, the masses of a proton and a neutron are given by:

$$m_{\text{p}} = 1.0078 \text{ u}$$

$$m_{\text{n}} = 1.0087 \text{ u}$$

The radioactive decay series discovered by Rutherford and the Curies can now be represented symbolically using our nuclear model. For example, if we start with a sample of radioactive $^{238}_{92}\text{U}$, the decay series begins with an alpha decay and transmutes into thorium-234:

$$^{238}_{92}\text{U} \rightarrow {}^{4}_{2}\text{He} + {}^{234}_{90}\text{Th}$$

The thorium isotope now decays by beta particle emission (in which a neutron is changed into a proton and an electron) and transmutes into the element protactinium-234:

$$^{234}_{90}\text{Th} \rightarrow {}^{0}_{-1}\text{e} + {}^{234}_{91}\text{Pa}$$

This decay series continues until a stable isotope of lead ($^{206}_{82}\text{Pb}$) is achieved. There are other decay series processes for different radioactive elements. *Artificial transmutation* can be achieved using nuclei bombardment. For example, the reaction $^{4}_{2}\text{He} + {}^{14}_{7}\text{N} \rightarrow {}^{1}_{1}\text{H} + {}^{17}_{8}\text{O}$ is achieved when alpha particles bombard nitrogen nuclei in a particle accelerator (using string magnetic fields).

In the early 1930s Irene Curie (1897–1956) and Frederick Joliot (1900–1958) discovered artificial radioactivity, which involved positron emissions (antielectrons, $^{0}_{1}\text{e}$). Antimatter had been predicted in the 1920s by theoretical physicist Paul Dirac (1902–1984). In 1932 James Chadwick (1891–1974) discovered the neutron (which 20 years earlier had been predicted to exist) by bombarding berylium nuclei with alpha particles: $^{4}_{2}\text{He} + {}^{9}_{4}\text{Be} \rightarrow {}^{12}_{6}\text{C} + {}^{1}_{0}\text{n}$.

With the discovery of the neutron and antimatter, physicists now began to think of matter and energy in terms of four fundamental forces:

1. Gravitation

2. Electromagnetism

3. Strong nuclear force (binds protons and neutrons together)

4. Weak nuclear force (explains radioactive decay)

The strong nuclear force has been used to explain nuclear binding energy (see the next section) as well as the quark model (which states that protons and neutrons are made up of smaller particles called quarks). The weak force was unified in the 1960s with electromagnetism (now called the electro-weak force) by Steven Weinberg (b. 1933), Sheldon Glashow

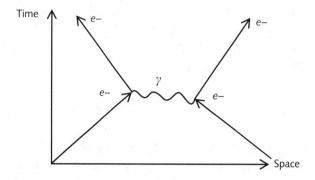

Figure 20.3 Typical Feynman diagram for electron-electron repulsion

(b. 1932), and Abdus Salam (1926–1996). Their work demonstrates that at high energy some of the fundamental forces do merge. These unifications (and their experimental evidence) have encouraged physicists to try and unify all four forces (and to question why there are four fundamental forces). So far, the goal of complete unification is elusive (as we stated with regard to the failure to merge quantum mechanics and a theory of gravitation).

With the discovery of cosmic rays and subatomic particles (such as mesons), theorists like Hideki Yukawa (1907–1981) began to think of the forces in terms of the exchange of particles. For example, Yukawa explained the strong force as a manifestation of the exchange of different kinds of mesons. The electromagnetic force can be explained by the exchange of photons. To simplify the complex mathematics underlying these theories, Richard Feynman (1918–1988) developed little diagrams in which the particle exchanges take place in both space and time (see Figure 20.3).

Nuclear Binding Energy

Physicists determining the masses of nuclei discovered that the actual mass was less than the sum of the constituent nucleons (protons plus neutrons). This missing mass (called the *mass defect*), accounted for by using Einstein's equation $E = mc^2$, is called the *binding energy* of the nucleus.

For example, consider the isotope ^4_2He that contains two protons plus two neutrons:

$$2m_p = 2(1.0078 \text{ u}) = 2.0156 \text{ u}$$

and

$$2m_n = 2(1.0087 \text{ u}) = 2.0174 \text{ u}$$

The total constituent mass should therefore be equal to $M = 4.033$ u. However, experiments show that the actual mass of the helium-4 nucleus is 4.0026 u. Therefore the mass defect, ΔM, is equal to 0.0304 u. This mass defect is proportional to the binding energy of the nucleus. For convenience, the energy equivalent of 1u is equal to 1.492×10^{-10} J (using $E = mc^2$) which is also equivalent to 931.49 MeV (which is the standard, accepted value when using $c = 2.9979 \times 10^8$ m/s). Remember that MeV refers to mega (or million) electron-volts. Thus:

BE (binding energy) = (931.49 MeV/u)(0.0304 u) = 28.3 MeV

We can also find the value for the *average binding energy per nucleon* by dividing the binding energy by the mass number. In this case:

$$BE_{avg} = 28.3 \text{ MeV}/4 = 7.08 \text{ MeV}$$

Problem   Calculate the average binding energy per nucleon for the tritium isotope of Hydrogen (^3_1H) given that the mass of tritium is 3.0160 u.

Solution

Step 1. We first need to identify the constituent nucleons and their masses:

1 proton = 1.0078 u

2 neutrons = 2(1.0087 u) = 2.0174 u

$M = 3.0252$ u

Step 2. Next, we find the mass defect by subtracting the total mass and the actual mass:

$\Delta M = 0.0092$ u

Step 3. The binding energy is now given by:

BE = (931.49 MeV/u)(0.0092 u) = 8.57 MeV

and the average binding energy per nucleon is equal to:

$BE_{avg} = 8.57 \text{ MeV}/3 = 2.86 \text{ MeV}$

Notice that the average binding energy per nucleon for tritium is less than the value for helium. This is an important observation since the average binding energy per nucleon is related to the stability of the nucleus. As we will learn, during *fusion* reactions hydrogen isotopes are fused into helium (this reaction takes place in the sun and other stars). Because the stability is increasing, energy is released. In a similar way, when uranium-235 undergoes *fission* the products have a higher average binding energy per nucleon than the reactants (and, again, energy is released).

Fission and Fusion

In 1939 Otto Hahn (1879–1968) and Fritz Strassmann (1902–1980) produced a splitting of the isotope $^{235}_{92}U$ by bombarding it with slow neutrons. The resulting reaction was analyzed by Lise Meitner (1878–1968) and Otto Frisch (1904–1979) and observed to be a fission reaction that released a huge amount of energy:

$$^{1}_{0}n + ^{235}_{92}U \rightarrow ^{92}_{36}Kr + ^{141}_{56}Ba + 3^{1}_{0}n + Energy$$

There are other fission reactions possible with uranium, but the main idea was that energy extracted from the process could be used as a source of energy. With World War II beginning, Albert Einstein wrote a letter to President Franklin Roosevelt urging him to pursue this research since Nazi Germany was known to be exploring the use of fission to make a bomb.

The secret Manhattan Project was begun under the direction of physicist J. Robert Oppenheimer (1904–1967), and on July 16, 1945, a fission-induced nuclear bomb was tested in New Mexico. On August 6, 1945, the first nuclear bomb was dropped on the Japanese city of Hiroshima. Several days later, a second bomb was dropped on the Japanese city of Nagasaki, ending World War II. After the war, fission power was used to develop power plants that would generate electricity. We will not go into the details (and controversies) of nuclear power plants, but it is a topic worth investigating for yourself.

The idea of a fusion of hydrogen into helium as a power source for the sun and stars was first proposed in the 1930s by physicist Hans Bethe (1906–2005). There are several ways in which hydrogen fusion can occur. A simplified common reaction sequence would proceed as follows:

$$^{1}_{1}H + ^{1}_{1}H \rightarrow ^{2}_{1}H + ^{0}_{1}e$$

$$^{1}_{1}H + ^{2}_{1}H \rightarrow ^{3}_{2}He$$

$$^{3}_{2}He + ^{3}_{2}He \rightarrow ^{4}_{2}He + 2^{1}_{1}H + Energy$$

The energy released in this fusion reaction is greater than the energy released in a fission reaction. On November 1, 1952, the United States tested the first hydrogen (thermonuclear) bomb in the Marshall Islands (Pacific Ocean) under the direction of physicist Edward Teller (1908–2003). Since that time, the world has struggled with nuclear proliferation as well as the need to find clean and renewable sources of energy. While there are many fission-powered nuclear power plants, scientists have not been successful with controlled fusion reactions.

From humble beginnings, the human race is now in possession of the knowledge to destroy itself or put itself on a path to peace and unlimited sources of clean, renewable energy. Nuclear, wind, solar, geothermal, and fossil-fuel power will compete as scientists, engineers, governments, and people struggle to move through a twenty-first century that Isaac Newton could never have dreamed of. Physics offers us a way to study how the universe works, but only humans can make the choices about how that knowledge is to be used.

Exercise 20.1

1. Calculate the mass-energy equivalence of an electron (m_e = 9.1 × 10^{-31} kg) in joules and MeV.

2. Using the Rydberg equation, calculate the wavelength of an ultraviolet photon in hydrogen Lyman series if $n = 3$.

3. Using the Bohr equation, calculate the frequency and wavelength of a photon emitted when an electron in a hydrogen atom makes a transition from level 5 to level 3.

4. Calculate the average binding energy per nucleon for the lithium isotope $^{7}_{3}\text{Li}$, which has an accepted atomic weight of 6.941 u.

5. The half-life of a radioactive substance is 2.5 hours. If 10 g of the substance remains after 12.5 hours, how much mass was present in the original sample?

Answer Key

Chapter 2   Units and Measurements

Exercise 2.1

1. a. 4 significant figures
   b. 3 significant figures
   c. 3 significant figures
   d. 3 significant figures

2. The sum is equal to 139.25 on a calculator. However, 3.4 has only 2 significant figures (the least), and so the answer should be reported as 140 (which may be ambiguous) or more precisely as 1.4×10^2 (2 significant figures).

3. The difference is equal to 56.3 on a calculator. Both numbers have 3 significant figures, so the answer can be reported as 56.3 (3 significant figures).

4. The product is equal to 21.93 on a calculator. However, 3.4 has only 2 significant figures, so the answer must be reported as 22 (2 significant figures).

5. The division is equal to 6.1. However, the number 2 has only one significant figure. Therefore, the answer must be reported as 6 (1 significant figure).

6. The squaring operation (4.5×4.5) is equal to 20.25 on a calculator. However, the number 4.5 has only 2 significant figures. The answer must be reported as 20. If the operation had been given as $(4.50)^2$, then the answer would have been reported as 20.3 (3 significant figures).

7. To find the average, you add all of the measurements and then divide by the number of measurements that you have. The division by the number of values is *not* the division of a measurement, so you do *not* have to consider how many

significant figures it has! Using a calculator, the average value appears as 3.583333. Since all of the measurements have 2 significant figures, the average value must be reported as 3.6 m.

Chapter 3   Graphical Analysis of Data

Exercise 3.1

1.

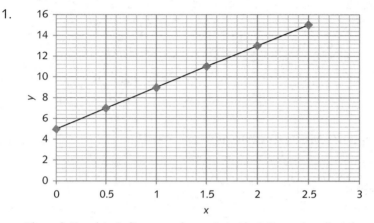

The relationship is linear and we state that "y varies directly with x." The slope of the line is equal to $k = \Delta y/\Delta x = 4$ and the y-intercept is equal to 5. Thus, the final equation is $y = 4x + 5$.

2.

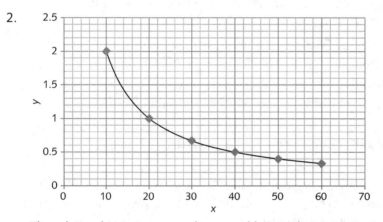

The relationship is inverse and we would state that "y varies inversely with x." Since the general equation is $xy = k$, from the data we can see that the value of k is 20. The final equation is $xy = 20$ or $y = 20/x$.

3.

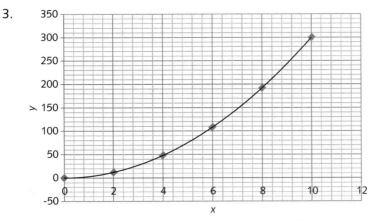

The relationship is quadratic and we would state that "y varies directly with the square of x." From the data, we can see that $k = y/x^2 = 3$. The final equation is $y = 3x^2$.

4.

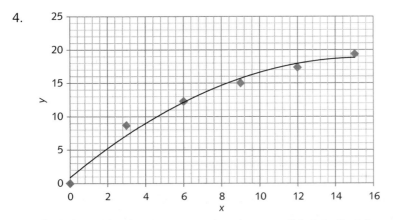

The relationship is a square root and we would state that "y varies directly with the square root of x." From the data, we can see that $k = y/\sqrt{x} = 5$. The final equation is $y = 5\sqrt{x}$.

Chapter 4   Linear Motion

Exercise 4.1

1. **d** = 10 km = 10,000 m and t = 2.5 hours = 9,000 seconds

$$\mathbf{v} = \frac{10,000 \text{ m}}{9,000 \text{ s}} = 1.110 \text{ m/s (west)}$$

2. $v_i = 25$ m/s and $v_f = 60$ m/s. Therefore, $\Delta v = 35$ m/s and $t = 7$ s.

   a. $a = \dfrac{v_f - v_i}{t} = \dfrac{60 \text{ m/s} - 25 \text{ m/s}}{7 \text{ s}} = +5 \text{ m/s}^2$

   b. $d = \left(\dfrac{v_i + v_f}{2}\right)t = \left(\dfrac{25.0 \text{ m/s} + 60.0 \text{ m/s}}{2}\right)t = (7 \text{ s}) = 297.5 \text{ m} = 298 \text{ m}$

3. $a = g = -9.8 \text{ m/s}^2$

    $d = -100$ m

    $v_i = 0$

    (Note that we need displacement to be equal to -100 m to avoid taking the square root of a negative number!)

   a. $d = \dfrac{1}{2}at^2 = \dfrac{1}{2}gt^2$

       $-100 \text{ m} = \dfrac{1}{2}\left(-9.8 \text{ m/s}^2\right)t^2$

       $t = 4.5$ s

   b. $v_f = v_i + at = 0 + (-9.8 \text{ m/s}^2)(4.52 \text{ s}) = -4.3 \text{ m/s (down)} = -44 \text{ m/s}$

4. $v_i = 30$ m/s

    $a = g = -9.8 \text{ m/s}^2$

    $t = 2$ s

   a. $d = v_i t + \dfrac{1}{2}at^2 = (30 \text{ m/s})(2 \text{ s}) + \dfrac{1}{2}\left(-9.8 \text{ m/s}^2\right)(2 \text{ s})^2$

       $= 60 \text{ m} - 19.6 \text{ m} = 40.4 \text{ m} = 40 \text{ m}$

   b. $v_f = v_i + at = v_i + gt = 30 \text{ m/s} + (-9.8 \text{ m/s}^2)(2 \text{ s}) = 10.4 \text{ m/s} = 10 \text{ m/s}$

   c. We can use $v_f^2 - v_i^2 = 2ad$. When $d = d_{max}$, $v_f = 0$, and $a = g = -9.8 \text{ m/s}^2$ (the vertical velocity is zero at the maximum height).

       $0 - (30 \text{ m/s})^2 = 2(-9.8 \text{ m/s}^2)d_{max}$

       $d_{max} = +45.92 \text{ m} = +46 \text{ m}$

5. a. The change in displacement is equal to the area under the velocity vs. time graph. This is a triangle with an area $= \dfrac{1}{2}(600 \text{ m/s})(60 \text{ s}) = 18,000 \text{ m}$.

    b. The average acceleration is equal to the slope of the line.

       $a = \dfrac{v_f - v_i}{t} = \dfrac{0 - 600 \text{ m/s}}{60 \text{ s}} = -10 \text{ m/s}^2$

6. a.

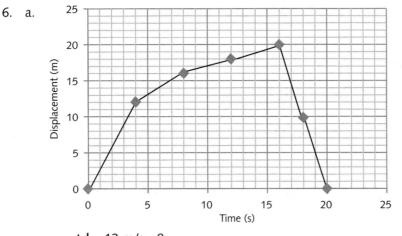

b. $\mathbf{v} = \dfrac{\Delta \mathbf{d}}{\Delta t} = \dfrac{12 \text{ m/s} - 0}{4 \text{ s}} = 3 \text{ m/s}$

c. The object travels a maximum distance of 20 m away from the origin and then travels 20 m back to the origin. The total distance traveled is therefore equal to 40 m.

d. The average speed is equal to the total distance divided by the total time.

$v = 40 \text{ m}/20 \text{ s} = 2 \text{ m/s}$

e. The total displacement of the object is zero since it returns to the origin. The average velocity is therefore equal to zero.

Chapter 5   Vectors

Exercise 5.1

1. A sketch (not drawn to actual scale) is shown:

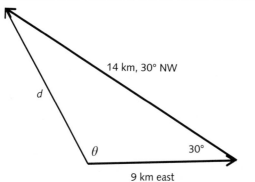

Solve for d using the law of cosines (notice that the interior angle is same as the given angle):

d^2 = (9 km)² + (14 km)² − 2(9 km)(14 km)cos(30°) = 81 km² + 196 km² −
(252 km²)(0.8660)

d^2 = 58.77 km² and, therefore, d = 7.67 km

To find the direction, we can solve for the angle θ and then associate it with the standard acute compass direction angles. Use the law of sines:

$$\frac{7.67 \text{ km}}{\sin(30)} = \frac{14 \text{ km}}{\sin \theta}$$

and a calculator yields θ = 65.87° NW (or 114.13° NE).

2. A sketch of the problem is shown:

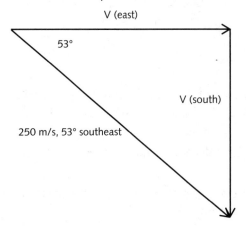

V (east)

53°

V (south)

250 m/s, 53° southeast

Now solve for each component using SOHCAHTOA (take south direction as *negative*):

v(east) = (250 m/s)cos(53°) = 150.45 m/s (east)

v(south) = −(250 m/s)sin(53°) = −199.66 m/s (south)

3. A sketch of the problem is shown:

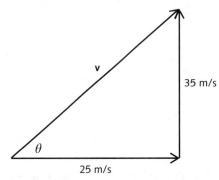

v

35 m/s

θ

25 m/s

Use the Pythagorean Theorem to solve for the launch velocity:

v² = (25 m/s)² + (35 m/s)² = 1850 m²/s²

v = 43.01 m/s

The direction angle can be found using the tangent function:

$\tan \theta = (35\ \text{m/s})/(25\ \text{m/s}) = 1.4$

and

$\theta = 54.46°$

4. A sketch of the situation is shown:

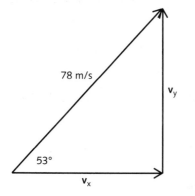

Use SOHCAHTOA to solve for the vertical and horizontal components of the given velocity:

$v_x = (78\ \text{m/s})\cos(53°) = 46.94\ \text{m/s (horizontal)}$

$v_y = (78\ \text{m/s})\sin(53°) = 62.29\ \text{m/s (vertical)}$

5. We have three vectors that need to be combined. If you think about this problem carefully, you can see that the 5-km east displacement and the 8-km west displacement simply combine to produce a resultant displacement of 3-km west (vectors in the opposite direction subtract). A sketch (and solution) of the reduced problem is shown (which is now simply a special 3, 4, 5 right triangle!):

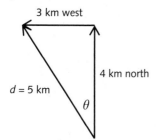

The direction angle can be found using the tangent function (which would be measured as west of north):

$\tan \theta = (3\ \text{km})/(4\ \text{km}) = 0.75$

and

$\theta = 36.87°$

Chapter 6   Forces

Exercise 6.1

1. The system is in equilibrium. That means that $\Sigma F_x = 0$ and $\Sigma F_y = 0$. The weight of the mass is $F_g = mg = (100 \text{ kg})(-9.8 \text{ m/s}^2) = -980 \text{ N}$.

    In the vertical direction, we have the vertical component of the tension equal (and opposite) to the weight of the mass:

    $\Sigma F_y = F_T \sin(30°) + (-980 \text{ N}) = 0$

    and, therefore,

    $F_T = 1960 \text{ N}$

2. The spring constant is equal to the slope of the line in a graph of Hooke's Law:

    a. $k = \Delta F / \Delta x = (100 \text{ N})/(0.5 \text{ m}) = 200 \text{ N/m}$

    b. $F = k\Delta x$ and we now know that $k = 200 \text{ N/m}$.

    $F = (200 \text{ N/m})(0.27 \text{ m}) = 54 \text{ N}$

3. a. Weight $= F_g = mg = (7 \text{ kg})(-9.8 \text{ m/s}^2) = -68.6 \text{ N}$

    b. $F_x = F \cos \theta = (50 \text{ N})\cos(40°) = 38.30 \text{ N}$

    $F_y = F \sin \theta = (50 \text{ N})\sin(40°) = 32.14 \text{ N}$

    c. To find the normal force, we observe that

    $\Sigma F_y = 0 = F_N + F_g + F_y = F_N + (-68.6 \text{ N}) + 32.14 \text{ N}$

    $F_N = 36.46 \text{ N}$

    d. $f_k = u_k F_N = (0.2)36.36 \text{ N} = 7.3 \text{ N}$ (to the left)

    e. $\Sigma F_x = ma_x$

    and

    $\Sigma F_x = F_x + f_k = 38.3 \text{ N} + (-7.3 \text{ N}) = 31.0 \text{ N}$

    $31.0 \text{ N} = (7 \text{ kg})a_x$

    $a_x = 4.43 \text{ m/s}^2$ (to the right)

4. For a mass sliding with no friction, we have uniform acceleration from rest. Therefore,

    $$d = \frac{1}{2}at^2$$

    and the acceleration can be calculated from this kinematics equation (choose down as negative):

    $$-1.2 \text{ m} = \frac{1}{2}a(2.3 \text{ s})^2$$

    $a = -0.45 \text{ m/s}^2$ (acceleration parallel to incline, a‖)

    Now, we know that F‖ = $mg \sin \theta$ and in this case (with no friction)

    F‖ = ma‖.

(2 kg)(−0.45 m/s²) = (2 kg)(−9.8 m/s²)sin θ

sin θ = 0.046

and, therefore,

θ = 2.6°

Chapter 7   Motion in a Plane

Exercise 7.1

1. a. The period of the motion is the time to complete one revolution. Thus:

    T = 20 rev/30 s = 0.67 s

    b. The magnitude of the tangential velocity is just the speed of the mass:

    v = distance/time = circumference/period = $2\pi R/T$

    $v = 2\pi(0.5$ m)/(0.67 s) = 4.69 m/s

    c. $F_c = mv^2/R$ = (0.2 kg)(4.69 m/s)²/(0.5 m) = 8.8 N

    d. When released, the mass becomes a horizontally launched projectile from a height of 2.2 meters with a velocity of 4.69 m/s. The time to fall can be found from $y = \frac{1}{2}gt^2$:

    −2.2 m = ½ (−9.8 m/s²)t^2

    t = 0.67 s

    and, therefore,

    $x = v_x t$ = (4.69 m/s)(0.67 s) = 3.14 m

2. To find the maximum height and the range, we first need to find the vertical and horizontal components of the launch velocity:

    v_{ix} = v cos θ = (125 m/s)cos(25°) = 113.29 m/s

    v_{iy} = v sin θ = (125 m/s)sin(25°) = 52.83 m/s

    Now, we can find t_{up}:

    $t_{up} = -v_{iy}/g$ = (52.83 m/s)/(9.8 m/s²) = 5.39 s = t_{down}

    Thus,

    $$y_{max} = \left|\frac{1}{2}gt_{down}^2\right| = \frac{1}{2}\left(9.8 \text{ m/s}^2\right)(5.39 \text{ s})^2 = 142.36 \text{ m}$$

    To find the range, we note that the total time is equal to 10.78 s and

    Range = $x_{max} = v_x t_{total}$ = (113.29 m/s)(10.78 s) = 1221.27 m

3. We are basically calibrating a pendulum clock. Solve the pendulum period equation for length:

    T = 1 s and g = 9.8 m/s² (magnitude)

    so

    $L = gT^2/4\pi^2$ = 0.25 m (25 cm)

4. In rotational equilibrium the sum of all torques must equal zero:

    $M_1gd_1 = M_2gd_2$

    $(1 \text{ kg})g(0.15 \text{ m}) = (0.4 \text{ kg})gd_2$

    $d_2 = 0.375 \text{ m}$

Chapter 8   Work, Energy, and Power

Exercise 8.1

1. a. The work done to lift the mass (work output) is equal to:

        $W_o = W_g = mgh = (15 \text{ kg})(9.8 \text{ m/s}^2)(1.25 \text{ m}) = 183.75 \text{ J}$

    b. IMA (ramp) = length/height = 2.5 m/1.25 m = 2

    c. Work done by the person (work input) is equal to:

        W_i = effort force × effort distance = (100 N)(2.5 m) = 250 J

    d. Work to overcome friction = $|W_i - W_o|$ = 250 J – 183.

        75 J = 66.25 J

    e. % efficiency = $(W_o/W_i) \times 100$ = (183.75 J/250 J) × 100 = 73.5%

2. a. First note that 40 kW = 40,000 W of power (that is 40,000 J/s). Also, 2 minutes is equal to 120 s. Since $P = W/t$, then

        Work = power × time = (40,000 J/s)(120 s) = 4,800,000 J

    b. Work = force × distance, so

        4,800,000 J = F(100 m)

        F = 48,000 N

3. To find the velocity of the cart on top of the 7-m hill (with no friction), we note that the total mechanical energy remains the same:

    $PE_i + KE_i = PE_f + KE_f$

    $$(100 \text{ kg})(9.8 \text{ m/s}^2)(12 \text{ m}) + \frac{1}{2}(100 \text{ kg})(5 \text{ m/s})^2 = (100 \text{ kg})(9.8 \text{ m/s}^2)(7 \text{ m})$$

    $$+ \frac{1}{2}(100 \text{ kg})v_f^2$$

    Now, solve for the velocity:

    11,760 J + 1,250 J = 6,860 J + 50v^2

    v = 11.1 m/s (forward)

4. The potential energy stored in the compressed spring will be converted into kinetic energy of the moving mass once it is released. This kinetic energy will then be

converted into gravitational potential energy. Since no mechanical energy is lost in all the conversions, we can simply state that:

PE_{spring} (before) $= PE_{mass}$ (after)

$$\frac{1}{2}kx^2 = mgh$$

$$\frac{1}{2}(20 \text{ N/m})(0.5 \text{ m})^2 = (0.5 \text{ kg})(9.8 \text{ m/s}^2)h$$

$h = 0.51$ m

Chapter 9   Momentum

Exercise 9.1

1. a.   Initial momentum $= m\mathbf{v}_i = (3 \text{ kg})(10 \text{ m/s}) = 30 \text{ kg·m/s}$

    b.   Total impulse = total area:

    From $t = 0$ s to $t = 10$ s, area $= 0$ N·s

    From $t = 10$ s to $t = 20$ s, area $= (10 \text{ N})(10 \text{ s}) = 100$ N·s

    From $t = 20$ s to $t = 35$ s, area $= (5 \text{ N})(15 \text{ s}) = 75$ N·s

    From $t = 35$ s to $t = 50$ s, area $= 0$ N·s

    Total impulse $= 175$ N·s $= 175$ kg·m/s

    c.   Impulse = change in momentum

    $175 \text{ kg·m/s} = (3 \text{ kg})(\mathbf{v}_f - 10 \text{ m/s})$

    $\mathbf{v}_f = 68.3$ m/s

2. a.   Momentum is conserved:

    $(4 \text{ kg})(2 \text{ m/s}) + (2 \text{ kg})(0 \text{ m/s}) = (4 \text{ kg})(0.66 \text{ m/s}) + (2 \text{ kg})\mathbf{v}_{2f}$

    $\mathbf{v}_{2f} = 2.68$ m/s

    b.   To test for an elastic collision, we need to see if the kinetic energy is conserved:

    $$KE_i = \frac{1}{2}(4 \text{ kg})(2 \text{ m/s})^2 = 8.0 \text{ J}$$

    $$KE_f = \frac{1}{2}(4 \text{ kg})(0.66 \text{ m/s})^2 + \frac{1}{2}(2 \text{ kg})(2.68 \text{ m/s})^2 = 8.05 \text{ m/s}$$

    Although the values for the kinetic energy (before and after) are not exactly equal, within experimental uncertainty, it appears that the collision can be considered elastic.

3. a.   Momentum is conserved and we need to consider the different velocities in terms of positive and negative:

    $(10 \text{ kg})(8 \text{ m/s}) + (5 \text{ kg})(-6 \text{ m/s}) = (15 \text{ kg})\mathbf{v}_f$

    $\mathbf{v}_f = 3.3$ m/s (right)

b. $KE_i = \frac{1}{2}(10\text{ kg})(8\text{ m/s})^2 + \frac{1}{2}(5\text{ kg})(-6\text{ m/s})^2 = 410\text{ J}$

$KE_f = \frac{1}{2}(15\text{ kg})(3.3\text{ m/s})^2 = 81.7\text{ J}$

Therefore, 328.3 J of KE was lost in the collision.

Chapter 10 Gravitation

Exercise 10.1

1. $F_g = (6.67 \times 10^{-11}\text{ N}\cdot\text{m}^2/\text{kg}^2)(2.5 \times 10^{10}\text{ kg})(4.5 \times 10^{12}\text{ kg})/(3.0 \times 10^4\text{ m})^2$

   $F_g = 8.34 \times 10^3\text{ N}$

2. The key to this problem is to remember that the distance needs to be from the center of the earth and measured in meters. If the altitude of the satellite is 500 km (500,000 m) above the surface of the earth, then the distance from the satellite to the center of the earth is:

   $R = R_e + h = 6.37 \times 10^6\text{ m} + 5 \times 10^5\text{ m} = 6.87 \times 10^6\text{ m}$

   $v_o = [(6.67 \times 10^{-11}\text{ N}\cdot\text{m}^2/\text{kg}^2)(5.98 \times 10^{24}\text{ kg})/(6.87 \times 10^6\text{ m})]^{1/2} = 7.6 \times 10^3\text{ m/s}$

   (For practice, recall that a *square root* is equivalent to an exponent of ½!)

3. $g = (6.67 \times 10^{-11}\text{ N}\cdot\text{m}^2/\text{kg}^2)(3.7 \times 10^{23}\text{ kg})/(5.5 \times 10^5\text{ m})^2 = 81.6\text{ m/s}^2$

   $F_g = mg = (10\text{ kg})(81.6\text{ m/s}^2) = 816\text{ N}$

4. The key to this problem is that the distance (assumed to be center-to-center) must be in kg and the time must be in seconds:

   $T = 1.77\text{ days} = 152,928\text{ seconds}$

   $R = 422,000\text{ km} = 422,000,000\text{ m} = 4.22 \times 10^8\text{ m}$

   Then

   $R^3/T^2 = (4.22 \times 10^8\text{ m})^3/(152,928\text{ s})^2 = (6.67 \times 10^{-11}\text{ N}\cdot\text{m}^2/\text{kg}^2)M_J/4\pi^2$

   $M_J = 1.9 \times 10^{27}\text{ kg}$

Chapter 11 Fluids

Exercise 11.1

1. We use the equation for static pressure: $P = \rho gh$, with $\rho_{water} = 1\text{ g/cm}^3 = 1000\text{ kg/m}^3$:

   $1.01 \times 10^5\text{ N/m}^2 = (1000\text{ kg/m}^3)(9.8\text{ m/s}^2)h$

   $h = 10.31\text{ m (about 34 US feet)}$

2. a. Archimedes' Principle states that a fully submerged object will displace a volume of water equal to its own volume. If the volume of the mass is

500 cm^3, then the volume of water displaced is equal to 500 cm^3. Because the density of water is 1 g/cm^3, the object will displace 500 g of water (0.5 kg).

b. The buoyant force is equal to the weight of the water displaced:

$F_B = F_g$ (water displaced) = mg = (0.5 kg)(9.8 m/s^2) = 4.9 N

c. To decide if the submerged object will float, we need to compare its weight in air with the buoyant force. Its mass is 200 g = 0.2 kg, therefore

F_g (object) = mg = (0.2 kg)(9.8 m/s^2) = 1.96 N

The object will float since the buoyant force is greater than the weight of the object in air.

3. a. The buoyant force will be equal to the difference between the weight of the object in air and the apparent weight of the object in water (when fully submerged):

F_B = 200 N − 50 N = 150 N

b. To find the volume of the object, we need to find the volume of the water displaced. The buoyant force is equal to the weight of the water displaced. The mass of water associated with 150 N is m = (150 N)/(9.8 m/s^2) = 15.31 kg = 15,310 g. Since the density of water is 1 g/cm^3, a displacement of 15,310 g corresponds to a volume of 15,310 cm^3 (which is also the volume of the object).

4. We use Pascal's formula:

(1000 N)/(30 m^2) = (50 N)/A_2

A_2 = 1.5 m^2

Chapter 12   Heat and the Kinetic Theory of Gases

Exercise 12.1

1. The specific heat of aluminum is c_{Al} = 0.9 J/g·°C. We use the conservation of heat:
(100 g)(0.9 J/g·°C)(200°C − T_f) = (400 g)(4.19 J/g·°C)(T_f − 20°C)
T_f = 29.2°C

2. a. $c_{solid} = \Delta Q/m\Delta T$ = (80 J)/(30 g)(160°C) = 0.167 J/g·°C
$c_{liquid} = \Delta Q/m\Delta T$ = (80 J)/(30 g)(160°C) = 0.167 J/g·°C
$c_{gas} = \Delta Q/m\Delta T$ = (60 J)/(30 g)(240°C) = 0.008 J/g·°C

b. $H_f = \Delta Q/m$ = (80 J)/(30 g) = 2.7 J/g
$H_v = \Delta Q/m$ = (100 J)/(30 g) = 3.3 J/g

3. (2 atm)(500 m^3)/(348 K) = (6 atm)(250 m^3)/T_2
T_2 = 522.6 K = 250°C

Chapter 13 Waves

Exercise 13.1

1. At 27°C, the velocity of sound is

    v = 331 m/s + (0.6 m/s)(27°C) = 347.2 m/s

    v_s = 45 m/s and f = 430 Hz

    therefore (the object is moving away from the observer),

    f' = (430 Hz)[(347.2 m/s)/(347.2 m/s + 45 m/s)] = 380.7 Hz

2. Nodes are produced every half-wavelength. If nodes occur every 35 cm, this means that

    λ = 70 cm = 0.70 m

    and

    f = 512 Hz

    $v = f\lambda$ = (512 Hz)(0.70 m) = 358.4 m/s

    Since the velocity of sound is 331 m/s at 0°C and increases by 0.6 m/s for each 1°C rise in temperature, we can see that Δv = 27.4 m/s and T = 45.7°C.

3. We know that $v = f\lambda$ and that $f = 1/T$, therefore $v = \lambda/T$. From this, we see that

    5 m/s = λ/(0.05 s)

    and, therefore,

    λ = 0.25 m.

Chapter 14 Static Electricity

Exercise 14.1

1. Charge B is attracted to both charge A and charge C. To find the net force on charge B, we need to first find the individual forces of attraction and then use vector methods (one-dimensional) to determine the net force. Use Coulomb's Law with appropriate units:

    $F_{A,B}$ = (9 × 10⁹ Nm²/C²)(7 × 10⁻⁶ C)(−2 × 10⁻⁶ C)/(0.5 m)² = −0.5 N
    (attractive to left)

    $F_{B,C}$ = (9 × 10⁹ Nm²/C²)(−2 × 10⁻⁶ C)(3 × 10⁻⁶ C)/(0.3 m)² = −0.6 N
    (attractive to the right)

    Now, the attractive force to the right (the force that charge C exerts on charge B) is *stronger* (by 0.1 N) than the attractive force that charge A exerts on charge B to the left. Thus,

    F_{net} (on charge B) = 0.1 N (to the right)

2. A proton is a positive elementary charge (q = +1.6 × 10^{-19} C). If the proton is accelerated from rest by a potential difference of 800 V, then ΔKE (eV) = 800 eV. Recall that ΔKE = Vq.

To convert from eV to J, we multiply the ΔKE by 1.6 × 10^{-19} J/eV, and, therefore, ΔKE = (800 eV)(1.6 × 10^{-19} J/eV) = 1.28 × 10^{-16} J.

Since the initial velocity is zero, we have

$$\Delta KE = \frac{1}{2}mv^2 = \frac{1}{2}\left(1.67 \times 10^{-27}\,kg\right)v^2 = 1.28 \times 10^{-16}\,J$$

And, therefore,

v = 3.9 × 10^5 m/s

3. We know that **E** = V/d, therefore **E** = (90 V)/(0.03 m) = 3000 V/m = 3000 N/C. To find the force, we use **F** = Eq = (3000 N/C)(1.6 × 10^{-19} C) = 4.8 × 10^{-16} N.

4. Charge sharing involves the motion of negative charges from a high concentration to a lower concentration until equilibrium is reached. The total charge for the system remains the same, but the charges will be distributed evenly. Thus Q_{total} = 2 C and the charge on each sphere will be equal to 1 C each.

Chapter 15   Electric Circuits

Exercise 15.1

1. We will use the equation:

$R = \rho L/A$

To find the cross-sectional area, we use

$A = \pi D^2/4 = \pi(0.0035\ m)^2/4 = 9.62 \times 10^{-6}\ m^2$

Thus,

$\rho = RA/L = (0.50\ \Omega)(9.62 \times 10^{-6}\ m^2)/(30\ m) = 1.6 \times 10^{-7}\ \Omega \cdot m$

2. The energy generated is equal to the work done (with time measured in seconds):

$W = VIt = (V^2/R)t = [(110\ V)^2/(1500\ \Omega)](300\ s) = 2420\ J$

3. This is a series circuit, therefore:

$R_{eq} = R_1 + R_2 + R_3 = 50\ \Omega + 30\ \Omega + 40\ \Omega = 120\ \Omega$

Since

$V_T = 60\ V,\ I_T = V_T/R_{eq} = 60\ V/120\ \Omega = 0.5\ A$

the total power generated is given by

$P_T = V_T I_T = (60\ V)(0.5\ A) = 30\ W$

Each voltage drop is given by:

V_1 = (50 Ω)(0.5 A) = 25 V, V_2 = (30 Ω)(0.5 A) = 15 V

and

V_3 = (40 Ω)(0.5 A) = 20 V

Notice that the total of the voltage drops is equal to 60 V, as expected.

4. In parallel, the resistors add reciprocally:

$1/R_{eq}$ = 1/30 Ω + 1/15 Ω + 1/10 Ω = 6/30 Ω

This means that

R_{eq} = 5 Ω

With a 20-V battery, the total current I_T = (20 V)/(5 Ω) = 4 A.

Finally, the voltage drop across each resistor is the same when we have a parallel circuit. Therefore,

I_1 = (20 V)/(30 Ω) = 0.67 A

I_2 = (20 V)/(15 Ω) = 1.33 A

and

I_3 = (20 V)/(10 Ω) = 2 A

Notice that the three currents add up to 4 amperes, as expected.

Chapter 16 Magnetism

Exercise 16.1

1. We use the equation $mv = Bqr$ to solve for the radius of the circular path:

    $r = mv/Bq$ = (9.1 × 10⁻³¹ kg)(4.5 × 10⁷ m/s)/(0.08 T)(1.6 × 10⁻¹⁹ C) = 3.2 × 10⁻³ m

    where we have used the magnitude of the charge of the electron.

2. We use the equation for finding the mass of an ion in a mass spectrometer:

    $m = B^2qr^2/2 V$ = (0.06 T)²(3.2 × 10⁻¹⁹ C)(0.75 m)²/2(700 V) = 4.63 × 10⁻²⁵ kg

3. We use $V_p/V_s = N_p/N_s$ so that

    $V_p = (N_p/N_s)V_s$ = (20,000/500)(110 V) = 4400 V

4. For the velocity selector:

    $v = E/B$ = (3000 N/C)/(0.002 T) = 1.5 × 10⁶ m/s

5. For electromagnetic induction:

    EMF = BLv = (4.0 T)(0.35 m)(15 m/s) = 21 V

    We use Ohm's Law to find the current:

    $I = EMF/R$ = (21 V)/(5 Ω) = 4.2 A

Chapter 17 Properties of Light

Exercise 17.1

1. Use Snell's Law:

   $(1.00)\sin(40°) = (2.42)\sin \theta_2$

   $\theta_2 = 15.4°$

   For the velocity of light:

   $v = c/n = (3 \times 10^8 \text{ m/s})/(2.42) = 1.23 \times 10^8 \text{ m/s}$

2. In air,

   $c = f\lambda$ and, therefore, $\lambda = c/f = (3 \times 10^8 \text{ m/s})/(5.0 \times 10^{14} \text{ Hz}) = 6.0 \times 10^{-7}$ m.

   In a diamond, the frequency does not change, but the velocity is (from problem 1)

   $v = 1.23 \times 10^8$ m/s

   and

   $v = f\lambda$

   Therefore, as the velocity changes the wavelength changes. This means that in the diamond:

   $\lambda = v/f = (1.23 \times 10^8 \text{ m/s})/(5.0 \times 10^{14} \text{ Hz}) = 2.26 \times 10^{-7}$ m.

3. For 700 lines/mm we have a slit separation distance of $d = 1.43 \times 10^{-6}$ m. We use the equation

   $x = \lambda L/d = (5.3 \times 10^{-7} \text{ m})(0.8 \text{ m})/(1.43 \times 10^{-6} \text{ m}) = 0.30$ m

4. One light-year is the distance that light travels in one year (in the vacuum of space). Thus, $c = 3 \times 10^8$ m/s and the number of seconds in one year is approximately $t = 31{,}557{,}600$ s. Since we treat the velocity of light as constant in space and $d = vt$ (from kinematics),

   d (one light-year) $= (3 \times 10^8 \text{ m/s})(31{,}557{,}600 \text{ s}) = 9.47 \times 10^{15}$ m

   To put this distance into context, it takes light about 8 minutes to travel the distance between the sun and the earth (approximately 93 million miles). One light-year is therefore approximately equal to 6 trillion miles and the nearest star to the sun (alpha Centauri) is 4.25 light-years away! It takes light over 4 years to travel from alpha Centauri to our eyes. In the northern hemisphere, the bright star in the night sky is called Sirius and it is about 8 light-years away. The Andromeda galaxy, however, is over 2 million light-years away. Consult an astronomy book for more fascinating details.

Chapter 18 Geometrical Optics

Exercise 18.1

1. To locate the focal point, we use two light rays. We know the positions of the object and the image, so we just follow our rules. Any light ray parallel to the principal axis refracts through the focal point, and any light ray drawn through the focal point refracts parallel to the principal axis. The situation is symmetrical and the constructed rays, identifying the two symmetrical focal points, are shown as a dashed set of rays. The required focal points are identified by lines and are symmetrical. A ruler is not needed.

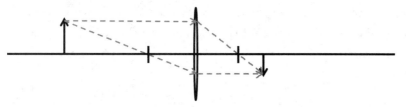

2. We recall that for a concave mirror, $C = 2f$. This means that $f = 5$ cm. Our object distance is given as $d_o = 8$ cm. Therefore:

    $$\frac{1}{5}\,cm = \frac{1}{8}\,cm + \frac{1}{d_i}$$

    and

    $d_i = 13.3$ cm (beyond point C as expected for case III)

    $S_i = (d_i/d_o)S_o = (13.3\ cm/8\ cm)(2\ cm) = 3.3$ cm

3. We use our rules for ray diagrams with a convex mirror (image should be virtual, smaller, and located behind the mirror). The virtual focal point is identified, so a ruler is not needed:

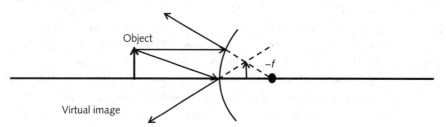

4. We can solve for the focal length using our equation:

    $$\frac{1}{f} = \frac{1}{8}\,cm + \frac{1}{12}\,cm$$

    and, therefore,

    $f = 4.8$ cm

5. In order to form parallel reflected rays (such as in a flashlight), the bulb should be placed at the focal point of the concave mirror.

Chapter 19   Quantum Theory of Light

Exercise 19.1

1. $E_{ph} = hf = (6.63 \times 10^{-34} \text{ J·s})(4.5 \times 10^{14} \text{ Hz}) = 2.98 \times 10^{-19} \text{ J} = 1.86 \text{ eV}$
   $W_o = hf_o = (6.63 \times 10^{-34} \text{ J·s})(1.5 \times 10^{14} \text{ Hz}) = 9.945 \times 10^{-20} \text{ J} = 0.62 \text{ eV}$
   $KE_{max} = E_{ph} - W_o = 1.9855 \times 10^{-19} \text{ J} = 1.24 \text{ eV}$

2. We know that
   $KE_{max} = V_o e^-$
   therefore, since
   $V_o = 1.5 \text{ V}$
   then
   $KE_{max} = 1.5 \text{ eV} = 2.4 \times 10^{-19} \text{ J}$
   $E_{ph} = hf = (6.63 \times 10^{-34} \text{ J·s})(5.8 \times 10^{14} \text{ Hz}) = 3.85 \times 10^{-19} \text{ J} = 2.4 \text{ eV}$
   Therefore,
   $W_o = E_{ph} - KE_{max} = 0.9 \text{ eV} = 1.44 \times 10^{-19} \text{ J}$

3. $E_{ph} = hf = hc/\lambda = (6.63 \times 10^{-34} \text{ J·s})(3 \times 10^8 \text{ m/s})/(5.65 \times 10^{-7} \text{ m}) = 3.52 \times 10^{-19} \text{ J}$

4. $\lambda = h/mv = (6.63 \times 10^{-34} \text{ J·s})/(1.67 \times 10^{-27} \text{ kg})(1.5 \times 10^8 \text{ m/s}) = 2.65 \times 10^{-15} \text{ m}$

5. $\mathbf{p} = h/\lambda$, which means that $\lambda = h/\mathbf{p} = (6.63 \times 10^{-34} \text{ J·s})/(7.5 \times 10^{-25} \text{ kg·m/s}) = 8.84 \times 10^{-10} \text{ m}$

Chapter 20   Atomic and Nuclear Physics

Exercise 20.1

1. We use $E = mc^2 = (9.1 \times 10^{-31} \text{ kg})(3 \times 10^8 \text{ m/s})^2 = 8.19 \times 10^{-14} \text{ J} = 5.12 \times 10^5 \text{ eV} = 0.512 \text{ MeV}$.

2. The Rydberg equation for hydrogen emission spectra is:
   $$\frac{1}{\lambda} = R\left(\frac{1}{1^2} - \frac{1}{n^2}\right)$$
   and in this problem $n = 3$. Make a simple substitution:
   $$\frac{1}{\lambda} = R\left(\frac{1}{1^2} - \frac{1}{3^2}\right) = (1.0968 \times 10^7 \text{m}^{-1})(0.8889) = 9.749 \times 10^6 \text{ m}^{-1}$$
   $\lambda = 1.026 \times 10^{-7} \text{ m (UV)}$

3. For the transition in a hydrogen atom from level 5 to level 3, we use the Bohr equation:

    $E_{ph} = E_i - E_f = E_5 - E_3 = (-0.54 \text{ eV}) - (-1.51 \text{ eV}) = 0.97 \text{ eV} = 1.552 \times 10^{-19} \text{ J} = hf$

    $f = E_{ph}/h = (1.552 \times 10^{-19} \text{ J})/(6.63 \times 10^{-34} \text{ J·s}) = 2.34 \times 10^{14} \text{ Hz}$

    $c = f\lambda$

    which means that

    $\lambda = c/f = (3 \times 10^8 \text{ m/s})/(2.34 \times 10^{14} \text{ Hz}) = 1.28 \times 10^{-6} \text{ m (IR)}$

4. The isotope ^7_3Li contains 3 protons and 4 neutrons (a total of 7 nucleons):

    $3m_p = 3(1.0078 \text{ u}) = 3.0234 \text{ u}$

    $4m_n = 4(1.0087) = 4.0348 \text{ u}$

    $M = 7.0582 \text{ u}$

    while the actual mass is given by

    $M_{Li} = 6.941 \text{ u}$

    The mass defect is

    $\Delta M = 0.1172 \text{ u}$

    and the binding energy is

    $BE = (931.49 \text{ MeV/u})(0.1172 \text{ u}) = 109.17 \text{ MeV}$

    The average binding energy per nucleon is therefore equal to 15.6 MeV.

5. The half-life formula is given by:

    $$m_f = \frac{m_i}{2^n}$$

    Since the half-life of the material is 2.5 hours, then a time interval of 12.5 hours is equal to five half-lives:

    $m_i = m_f 2^n = (10 \text{ g})(2^5) = 320 \text{ g}$

Index

Note: *f* denotes figure; *t*, table